Mimesis and Science

Studies in Violence, Mimesis, and Culture Series

Mimesis and Science

Science

Empirical Research on Imitation and the
Mimetic Theory of Culture and Religion

Edited by Scott R. Garrels

Michigan State University Press · *East Lansing*

Copyright © 2011 by Michigan State University

♾ The paper used in this publication meets the minimum requirements of ANSI/
NISO Z39.48-1992 (R 1997) (Permanence of Paper).

Michigan State University Press
East Lansing, Michigan 48823-5245

Printed and bound in the United States of America.

17 16 15 14 13 12 11   1 2 3 4 5 6 7 8 9 10

LIBRARY OF CONGRESS CATALOGING-IN-PUBLICATION DATA

Mimesis and science : empirical research on imitation and the mimetic theory of
    culture and religion / edited by Scott R. Garrels.
        p. cm. — (Studies in violence, mimesis, and culture)
    Includes bibliographical references and index.
    ISBN 978-1-61186-023-8 (pbk. : alk. paper)
1. Imitation.   I. Garrels, Scott R.   II. Title.   III. Series.
BF357.M546 2011
302—dc22
                                                            2011008834

Cover design by Sharp Des!gns, Inc.
Book design by Aptara, Inc.
Cover art is M.C. Escher's "Encounter" © 2011 The M.C. Escher Company-Holland.
All rights reserved. www.mcescher.com

g green Michigan State University Press is a member of the Green Press
  press
  INITIATIVE Initiative and is committed to developing and encouraging eco-
logically responsible publishing practices. For more information about the Green
Press Initiative and the use of recycled paper in book publishing, please visit
www.greenpressinitiative.org.

Visit Michigan State University Press on the World Wide Web at:
www.msupress.msu.edu

This book is dedicated to my wife Joy, who graciously labored
with me on this project through our engagement,
marriage, and the first year
of our daughter's life.

Contents

Preface

In May 2002, an interdisciplinary group of scientists and philosophers gathered in Royaumont Abbey, France, for an unprecedented meeting of the "imitative minds." The purpose of this four-day conference was to discuss an array of recent scientific breakthroughs concerning the enormous, yet profoundly overlooked role of imitation in human life. The disciplines represented were as diverse as child development, primatology, neurophysiology, social psychology, and philosophy of mind. The conference proceedings, which included nearly 50 contributors, were later published in a two-volume set that has since become the definitive collection on contemporary imitation theory and research[1]—with one major exception.

Decades before advances in the empirical sciences prompted a surge of interest in imitation, the French literary analyst and cultural theorist René Girard had already developed his own theory of imitation, or *mimetic theory*. Beginning in the early 1960s, Girard identified the generative role of imitation in human desire and motivation,[2] including a profound understanding of the relationship between imitation, violence, and the origin and structure of human culture and religion.[3] In fact, in March 2005 (the same month and year that the above volume on the science of imitation was published) Girard was inaugurated into the French Academy, an honor bestowed on him as a result of nearly 50 years of writing and research devoted to his overall argument that our great capacity for imitation is far and away the most important, yet least understood of all human abilities. It is therefore remarkable that the respective literatures on Girard's mimetic theory, and the contemporary experimental research on human imitation have remained largely compart-mentalized, with very little reference to, or even apparent awareness of, one another.

Mimesis and Science, therefore, represents our initial attempt at estab-lishing a relationship between these two groundbreaking fields of research. In doing so, we have brought together, for the very first time, some of the foremost scholars of Girard's mimetic theory with leading imitation research-ers in the cognitive and social sciences. Among the authors are many who have pioneered research in their own field and who have produced seminal works on imitation as a result. Additional authors are included whose work and disciplines intersect significantly with the implications of these areas of research. While not exhaustive, this volume provides an overview of some of the major findings and interpretations concerning human imitation across the diverse disciplines of developmental psychology, cognitive neuroscience, neurophysiology, comparative psychology, psychiatry, anthropology, literary analysis, and philosophy. Core themes include the ubiquitous, foundational, and paradoxical nature of imitation in human life, including its essential role in human development, cognition, culture, and religion. To be sure, there are many questions and areas of research that are left unanswered and unexplored. Regardless, our overall goal is to help initiate the process of cross-fertilization between mimetic scholars and empirical researchers so that significant links can continue to develop, encouraging new thinking, creative scholarship, future experimental research, and a more profound and encompassing view of the role of imitation in human life.

The inspiration for this project came about while I was a doctoral student in clinical psychology at Fuller Theological Seminary in Pasadena, California. While at Fuller, I came across Girard's mimetic theory and the contemporary experimental research on human imitation almost simultaneously. My intro-duction to mimetic theory came from one of my mentors, Jim Steinwedell, who recommended a book based on Girard's work after my repeated ques-tioning about the nature of violence and religion in the modern world.[4] My introduction to the experimental research came soon thereafter when my advisor, Warren Brown, presented an article to our neuropsychology research lab on the recent discovery of mirror neurons.[5] This article led me further to explore what the empirical sciences had to say about human imitation. The more I became acquainted with Girard's literary and anthropological theory, alongside the contemporary research on imitation in the sciences, the more compelling the subject of human imitation became and, to my great surprise, the more I found that practically nothing had been done to integrate these overlapping areas of research. After publishing an initial article of my own, I concluded that bringing these scholars and researchers together themselves would be the best way to further this work.[6]

This project was made possible by a grant from the Metanexus Institute as a Templeton Advanced Research Program. The Girardian foundation, Imitatio: Integrating the Human Sciences later provided additional funding. Together, these grants allowed the authors of this volume to meet for three different two-day symposia held at Stanford University and at Le Collège des Bernardins and the École Normale Supérieure in Paris, France, in 2007 and 2008 in order to discuss their respective disciplines, present relevant research, and engage in extensive dialogue, commentary, and criticism with scholars and researchers in other disciplines. This series of workshops was an integral part of this unique and unprecedented collaboration, especially since many of the scholars and researchers were learning about each other's work for the first time. The extensive time spent together allowed the authors to work through many of the challenges inherent in interdisciplinary dialogue in order to reach a level of cohesiveness that is too often lacking in collective works of this kind.

This book, and the meetings that it represents, was a collective endeavor from beginning to end. As result, I would like to thank the many individuals and organizations that contributed to making it possible. From Fuller's Graduate School of Psychology and the Travis Research Institute at Fuller: Nancey Murphy, Lynn Paul-Sternberg, Winston Gooden, James Van Slyke, Elke Rechberger, Al Dueck, Alexis Abernethy, Marta Cenac-Mehedinti, and Chris Waters. Special thanks to Warren Brown for his enthusiasm and support as director of the Travis Research Institute, his participation in our group meetings, and his assistance with editing several portions of this book; and to James Steinwedell for his role and encouragement in getting me both into, and through, this incredible journey.

From the Colloquium on Violence and Religion: James Alison, Pierpaolo Antonello (St. John's College, Cambridge), Paul Gifford (University of St. Andrews), Martha Rieneke (University of Northern Iowa), and Gil Bailie (The Cornerstone Forum) for their participation and/or encouragement at various points in this project. Special thanks to Robert Hamerton-Kelly (Stanford University) and Andrew McKenna (Loyola University of Chicago), who early on encouraged the formation of this project, and later, through Hamerton-Kelly's creation of Imitatio, helped keep it afloat; to Trevor Merrill (University of California, Los Angeles) for his friendship and invaluable support, both intellectually and practically, at every stage of this work; and especially to

William Johnsen at Michigan State University for his patience and scholarly dedication to this book.

I would also like to thank the following persons and institutions for their financial support and/or use of their facilities: William Grassie and Elizabeth Kenny at Metanexus Institute; Peter Thiel, Lindy Fishburne, and Amber Fowler at Imitatio; Byron Bland and the Martin Daniel Gould Center for Conflict Resolution at Stanford University; Benoît Chantre at Le Collège des Bernardins and the Association Recherches Mimetiques; and the École Normale Supérieure.

Finally, I am deeply grateful to each of the authors of this volume for their willingness to participate in this project, and for the considerable amount of time and work they put into the project meetings and their resulting chapters. I can only hope that they learned as much from each other throughout this process as I was able to learn from them.

Lastly, and most importantly, I would like to thank René and Martha Girard for their immense generosity of time and personal support. René's work was the main source of inspiration for this collaboration, and his participation with our group at each of the meetings was a genuine gift to us all.

NOTES

1. Susan Hurley and Nick Chater, eds., *Perspectives on Human Imitation: From Neuroscience to Social Science*, vol. 1, *Mechanisms of Imitation and Imitation in Animals*; vol. 2: *Imitation, Human Development, and Culture* (Cambridge, MA: MIT Press, 2005).

2. René Girard, *Deceit, Desire, and the Novel: Self and Other in Literary Structure*, trans. Yvonne Freccero (Baltimore: Johns Hopkins University Press, 1965).

3. René Girard, *Violence and the Sacred*, trans. Patrick Gregory (Baltimore: Johns Hopkins University Press, 1977).

4. Gil Bailie, *Violence Unveiled: Humanity at the Crossroads* (New York: Crossroad, 1995).

5. Vittorio Gallese et al., "Action Recognition in the Premotor Cortex," *Brain* 119 (1996): 593–609; Giacomo Rizzolatti, Luciano Fadiga, Vittorio Gallese, and Leonardo Fogassi, "Premotor Cortex and the Recognition of Motor Actions," *Cognitive Brain Research* 3 (1996): 131–41.

6. Scott R. Garrels, "Imitation, Mirror Neurons, and Mimetic Desire: Convergent Support for the Work of René Girard," *Contagion: Journal of Violence, Mimesis, and Culture* 12–13 (2006): 47–86.

CHAPTER 1

Human Imitation: Historical, Philosophical, and Scientific Perspectives

Scott R. Garrels

In the science of man and culture today there is a unilateral swerve away from anything that could be called mimicry, imitation, or mimesis. And yet there is nothing, or next to nothing, in human behavior that is not learned, and all learning is based on imitation. If human beings suddenly ceased imitating, all forms of culture would vanish. . . . To develop a science of man it is necessary to compare human imitation with animal mimicry, and to specify the properly human modalities of mimetic behavior, if they indeed exist.

—René Girard, *Things Hidden since the Foundation of the World*

Few areas of recent research have shed as much light on our understanding of human nature as those that address with fresh insight the unique and foundational properties of human imitation. Far from being the simple and mindless act that we typically associate it with ("monkey see, monkey do"), imitation is now understood as a complex, generative, and multidimensional phenomenon at the heart of what makes us human. In fact, imitation may very well be the basis for not only how we learn, but also how we understand each other's intentions and desires, establish relational bonds, fall in love, become jealous, compete with one another, and violently destroy each other, all the while operating largely outside of our conscious awareness.

This renewed view of imitation as one of the most compelling and overlooked capacities of the human species can be seen across a wide range of disciplines, in both the humanities and the cognitive and social sciences. While divergent in their methodological goals and aims, each approach shares in common the view that our pervasive tendency to imitate highlights the essentially *relational* nature of the human mind and person—a perspective that is antithetical to what has been the predominant and overarching view of Western philosophical and scientific thinking for over the last four hundred years; namely, that humans are fundamentally self-enlightened or *autonomous* beings. As a result, contemporary research on human imitation offers new and exciting potential for exploring and reevaluating fundamental questions about what it means to be human.

However, despite recent work attesting to its essential role, imitation still remains largely misunderstood within many fields as a secondary, rather than central, feature of human life. This is due largely to many enduring and deeply ingrained philosophical and conceptual biases concerning imitation, some dating back thousands of years, which only recently have come under critical question and reevaluation. In addition, there are still too few scholars investigating the obvious links between recent research on imitation and existing theories of human development, cognition, and culture. What is more, the present need for interdisciplinary collaboration addressed by this volume comes from the remarkable fact that many scholars and researchers have arrived at their groundbreaking conclusions concerning human imitation in complete isolation from one another. Nowhere is this more poignantly demonstrated than in the separate developments of René Girard's mimetic theory, and contemporary experimental research on human imitation.

Beginning with his literary research in the early 1960s, Girard discovered certain repetitive patterns of imitation that proved reliable and consistent throughout different periods of history. He concluded that humans not only imitated surface behaviors but also, more fundamentally, were motivated, unknowingly, according to imitative or *mimetic* principles of desire. *Mimetic desire*, as Girard perceived it, was an essential yet paradoxical feature of human relations that generated social bonding and affiliation as well as intense forms of competition, rivalry, and even violence. From this understanding, Girard went on to interpret a wide range of anthropological texts and phenomena. He further concluded not only that imitation was essentially human, but that it was *the* founding social force that propelled proto-humans to establish culture and religion through primitive forms of ritual sacrifice—a social mechanism that contained escalating in-group violence by deflecting it against an outside or surrogate victim. In effect, Girard developed an entire theory of human intersubjectivity and religious and cultural evolution on the basis of his understanding of imitation. As a result of his extensive writing and research over the last 50 years, it is "clear to a wide range of scholars that Girard is one of the most original and influential cultural theorists on the contemporary scene."[1]

Until recently, such a pervasive and foundational role of imitation in human life was either largely ignored or misunderstood by experimental scientists. This is no longer the case. Within the last few decades, there has been a dramatic surge of interest across a wide range of disciplines. Researchers now argue that imitation is an innate, and characteristically human, ability that guides cognitive and social development from the very beginning of life, both from developmental and evolutionary perspectives. Not only does imitation function powerfully in the parent-infant dyad to bring about experience-dependent neurocognitive growth, but it thrives in adulthood as one of the most ubiquitous, organizing, and unconscious characteristics of human social relations. In fact, so foundational is our capacity to imitate, that many researchers believe it to be the linchpin that contributed to a wide-scale neural reorganization of the brain, allowing for the coevolution of more complex social, cultural, and representational abilities from earlier primates to humans. Undoubtedly, empirical investigations on human imitation are now among the most innovative areas of research contributing to the future of a more unified and coherent understanding of the cognitive and social sciences.

The parallels between Girard's literary and anthropological insights and the conclusions based on contemporary empirical investigations are therefore remarkable and deserve a more detailed and comparative review. Yet surprisingly, despite the inherent overlap between these two fields, there have been no substantial inquiries concerning the critical question of human imitation that have attempted to apply and synthesize recent empirical research with Girard's mimetic theory. The purpose of this book, then, is to help initiate such a process by pioneering collaboration between scholars of Girard's mimetic theory and experimental researchers in the cognitive and social sciences.

Is imitation unique to the human species? If so, how is such a seemingly mundane ability central to the generative and complex processes necessary for human development and religious and cultural evolution? Furthermore, is it possible that imitation is foundational to our most prized forms of human intelligence, empathy, and sociality *as well as* our most enduring and reproachful acts of human conflict, violence, and destruction? If so, how do we understand this paradox? And why is it so difficult to acknowledge and come to terms with our natural human tendency to imitate?

These are some of the main questions addressed in this book. In doing so, we have put together a collection of chapters that display the unique theoretical links that can now be made from the neural basis of social interaction to the structure and evolution of human culture and religion. These chapters also demonstrate the valuable contribution that contemporary experimental science can make to Girard's theoretical formulations by clarifying important cognitive, developmental, and evolutionary dimensions of imitation that are not accounted for in his mimetic theory—and in complementary fashion, how empirical researchers may benefit from previously unsuspected implications of human imitative phenomena already discovered by Girard and other mimetic scholars working from a more anthropological perspective. As this volume reveals, when taken together, both imitation research and mimetic theory provide a complementary set of theories that lend greater clarity and explanatory depth to the study of human imitation than is found in either field alone.

Ultimately, we feel that furthering this collaborative effort is essential, not only for the impact it might have on a wide range of disciplines in arriving at a more encompassing view of our human imitative nature, but also for the potential of this shared knowledge to shed light on some of the most pressing and complex questions of our contemporary world; namely, how are we to understand the immense quandary of human violence (even the threat of violence) as manifest in a variety of forms on the world scene, and what are the religious, cultural, and relational factors that contribute to its proliferation or reduction? Mimetic scholars and imitation researchers are on the cutting edge of some of the most innovative breakthroughs in understanding our distinctive capacities for both incredible acts of empathy and compassion as well as mass antipathy and violence. If genuine dialogue can be created and sustained between these two bodies of research, then perhaps we may benefit from a greater appreciation of the incredible nature of human life, culture, and religion—one that may very well be essential to understanding how human relationships can be transformed through infinitely more imaginative and nonviolent ways of relating. This, at least, is what we hope to convey and inspire with the following chapters.

The purpose of this introductory chapter is to provide some background to the theories and issues considered in this book. A brief historical sketch will serve to highlight the significance of the more recent developments of Girard's mimetic theory and the new science of human imitation, as well as the need for integration between the two.

A Brief History of Imitation

Contemporary theory and research produces a profound curiosity about past conceptualizations concerning human imitation. If it is true that imitation is one, if not *the* fundamental mechanism driving human development, motivation, and cultural evolution, then how have previous generations of philosophers and social scientists understood this central human ability? One would think that if imitation were so vital and important to human life, then it would be evident and known in some ways by the experiences and observations of it throughout history. Indeed, imitation has been central to philosophical discourse since the foundation of Western civilization. The issue at hand is not so much that imitation has been unrecognized, but rather the way in which it has been treated and emphasized during different periods of history, including important dimensions that, until recently, have been overlooked.[2]

From Antiquity through the Renaissance

Beginning in ancient Greece and extending through the Renaissance, imitation in human life was conceptualized primarily in two ways: the imitation, or *re-presentation* of nature and human behavior in the arts, and the imitation of preceding artists, writers, religious leaders, philosophers, and tradesmen alike.[3] These two dimensions reflect the most self-evident, or consciously recognizable, features of human imitation that are still emphasized today—that is, our unique representational skills and our unsurpassed capacity for cultural learning, including the creation, transmission, and improvement upon knowledge across generations.

Plato and Aristotle were the first to treat the subject of *mimesis* (the Greek word for imitation) at length, with each representing opposing attitudes that would prove influential throughout history, including up to the present day. Plato's ideas are considered a primary source of our modern tendency to devalue imitation. In Plato's system of thought, all elements of *matter* in the created world were a mere shadow, or imitation, of their absolute *Form* in the nonmaterial realm of the Good. In his pursuit of objective truth and ethical grounds for the moral life, Plato believed that reason alone was capable of perceiving directly these most absolute forms of reality. On the other hand, knowledge based on mimesis in the arts was considered a devalued form of the truth since it was based on a twice-removed imitation of reality (e.g., the artist's painting of a table was a copy of the material object, which itself was a copy of the absolute form). Plato also perceived that the arts played on the emotions of their audience in a way that exaggerated certain aspects of reality for their effect (e.g., plays that dramatized the life of the gods and their relation to humans).

While open at times to the importance of mimesis in the arts, Plato "saw in philosophy a much more potent and secure pedagogy, whereby reason, unshackled by feeling, passion, and the other agencies of matter, might more effectively form and liberate the human spirit."[4] Thus, in Plato's idealistic system of knowledge, mimesis was viewed as a weaker version and potentially grave distortion of the truth, one that could not measure up to the percept of abstract reason. Indeed, "mimesis was seen to limit intelligence, destroy identity, and even lead to murder or suicide!"[5] As a result, the Platonic tradition has been viewed, in part, as having an "insidious influence" on our contemporary attitude toward imitation. "Even today, conformism, lack of initiative, and submissiveness are associated with imitation."[6]

Aristotle had a much more positive view of imitation. Early on, he affirmed that "Imitation is natural to man from childhood, one of his advantages over the lower animals being this, that he is the most imitative creature in the world, and learns at first by imitation."[7] His *Poetics* dealt primarily with drama and art as the mirror of nature with respect to the natural world and human behavior. In contrast to Plato's idealized separation between form and matter, Aristotle saw the two as fundamentally interdependent, where "form needed matter to express its full potential and grow to maturity; and matter needed form for determination and stability. . . . The subsequent intentional union of the mind with the world about it was direct, in the sense that things were known in themselves and not merely as shadows; hence poetry, a special kind of imitative knowledge, was proportionately direct, unlike Plato's imitation of an imitation."[8] From this perspective, the practice of mimesis was seen in all of its positive aspects and benefits for furthering knowledge and creating beauty that both strove to represent, and in some respects improve upon, the natural world, including human conduct and knowledge in general.

The Aristotelian view of imitation dominated discourse from antiquity through the Renaissance period. In fact, the wide range of cultural and artistic innovations characteristic of the Renaissance were based philosophically on the importance of imitation, particularly the communal aspects of artistic collaboration and the imitation of previous generations. "Despite the various ways ancient authors cast their discussions of imitation, all agreed that it is inevitable, and desirable to the extent that the imitator recasts his source and appropriates it to his own inventive capacity; only in this way can the art evolve and avoid decline."[9] This "mimetic tradition" was not only applied to the arts, but was also the basis for continued advancement and reform in religious, historical, and philosophical traditions.[10] Thomas à Kempis's *Imitation of Christ* in the fifteenth century is a classic example of the positive attitude toward imitation that was widely evident throughout this period.[11] In effect, in contrast to our modern notion of limiting growth, or mere conformity, imitation was seen as a primary means of social and cultural progress throughout this long period of human history. "Undoubtedly the change in attitude in modern times has made more difficult our understanding of imitation and our capacity to perceive its benefits and its ties to Renaissance inventiveness."[12]

It is therefore ironic that the figure who would become known as the quintessential "Renaissance Man" was also the one most opposed to the mimetic tradition. According to Ackerman, "Leonardo da Vinci was the only writer who disapproved of *all* imitation in the classical sense. He wrote of it: 'No one should ever imitate the *maniera* of another because he will be called a nephew and not a child of nature with regard to art. Because things in nature exist in such abundance, we need and we ought rather to have recourse to nature than to those masters who have learned from her.'"[13] It is hard to imagine, however, that da Vinci was not significantly influenced by others in the development of his art, and even more difficult to imagine, given his substantial influence, that subsequent artists did not then imitate him in his dictate to not imitate! Indeed, the generation following da Vinci would witness a virtual end of the mimetic tradition.[14] The attempt to rely primarily on oneself, and therefore deny imitation, would find its fullest expression in modern philosophy.

Modern Philosophy

In contrast to the mimetic tradition, which looked primarily to community, both past and present, as a foundational source of knowledge, modern philosophy found its grounding in the idea of an autonomous self. This core belief emphasized the importance of individual strivings over and against the role and function of social influences in the construction of knowledge and the operation of the human mind.

René Descartes is generally considered the principal figure who, with his mid-seventeenth-century philosophical credo "I think, therefore I am," gave a defining voice to modern philosophy and the Enlightenment way of thinking. Stephen Gardner's description of Cartesian rationality is worth repeating at length:

> In Descartes case . . . every individual, every ego cogito, is in principle empowered to judge the truth absolutely for himself—meaning, he not only must make a judgment but has all the resources necessary to make it rightly, independently of nature, tradition, other humans, revelation, or the senses—provided that he has the strength of mind or "resoluteness" demanded to achieve metaphysical certainty. In Descartes' eyes, the certainty of the ego is both necessary and sufficient to establish absolute truths. This is a decisive source of Enlightenment in its radical forms, which invoke the autonomy of human knowledge (and so of will).[15]
>
> Thus Descartes creates the modern myth of autonomy in its first explicit shape—the belief that the individual has the resources to determine the object of his desire without support of other(s) or imitation, a self capable of relying upon itself alone, without external props. The individual can will his own self in a radically independent way. Even if for reasons of prudence he must outwardly conform himself to the laws and manners of his country, inwardly he retains the independence of his true self. The only question is, whether an individual has the "resoluteness" (now consistently elevated to the chief virtue) to adhere to the dictate of his own reason.[16]

In many ways, Descartes's position reified Plato's dualistic view of reality that presumed rational thought could be separated out from the influence of mimesis and emotions. This "Cartesian split" between perception/emotion and thought/behavior brought with it the erroneous belief that each individual could consciously ward off all human passions and social influences in the pursuit of a more objective truth.[17] In effect, Descartes implicitly denied that imitation was an essential and ongoing feature of human cognition, motivation, and behavior. In doing so, he was influential in setting up a long-standing dichotomy in the cognitive and social sciences between perception and action systems.[18] Not only that, but the concept of imitation, or mimesis, itself seemed all but explained away or devalued as an important idea worth continued philosophical exploration.[19]

The impetus for modern philosophy's view of the self as autonomous has been interpreted in part as a response to the rampant violence that spread throughout Europe as a result of the Thirty Years' War.[20] Religious and political conflict, beginning with the Catholic and Protestant churches, led to one of the most violent and enveloping wars in all of European history. As a result, many traditional sources of knowledge and authority were brought into question and eventually replaced by philosophical rationalism and scientific empiricism. In effect, the mimetic tradition was essentially overturned by what many hoped would prove to be a more objective perspective from which to bring about a more peaceful and prosperous future. The failure of such an "Enlightenment project" based on the premise of a purely autonomous and rational self has been the subject of much postmodern criticism, and is also evidenced by the unprecedented level of collective violence seen throughout the twentieth century.

Clearly, the origins and development of modern philosophy are complex and extend far beyond the writings and influence of Descartes, including the fact that many notable philosophers in the modern period did not reflect the Enlightenment spirit.[21] Nonetheless, it is difficult to overestimate the overall effect that the Enlightenment tradition had on modern philosophy and the history of Western thought for over the last four hundred years. The social sciences and medical models developed during this time were so influenced by an autonomous view of the human mind and person, that only in the past 50 years have we begun to make incremental shifts in our conceptual paradigms, allowing for more diverse perspectives across a wide range of disciplines.[22]

Modern Psychology

The end of the nineteenth century is generally considered the period when the discipline of psychology was born as a separate branch of the sciences. Despite the clear dominance of Enlightenment thinking at the time, interest in imitation was revived for a short period, and even viewed by several prominent thinkers as *the* defining human characteristic. However, by the turn of the century, this emphasis would all but disappear as imitation was substantially marginalized by the most influential theories of modern psychology.

The French social psychologist Gabriel Tarde is perhaps the most notable imitation theorist of this early period in modern psychology. In 1890, Tarde published his *Laws of Imitation*, where he defines imitation as constituting "every impression of an inter-psychical photography, so to speak, willed or not willed, passive or active. If we observe that wherever there is a social relation between two living beings, there we have imitation in this sense of the word."[23] For Tarde, imitation was the essential "social fact" to which all other social processes, including the most complex and innovative forms of behavior, ideas, and culture, could be related. As a result, Tarde's views on imitation directly challenged the idea of an autonomous self. According to Tarde, "We err in flattering ourselves that we have become less credulous and docile, less imitative, in short, than our ancestors. This is a fallacy, and we shall have to rid ourselves of it." In fact, Tarde went so far as to compare human sociality to a state of hypnosis: "The social like the hypnotic state, is only a form of dream, a dream of command and a dream of action. Both the somnambulist and the social man are possessed by the illusion that their ideas, all of which have been suggested to them, are spontaneous."[24] As a result of the work of Tarde, as well as many other notable authors, the subject of imitation was widely discussed and debated by the end of the nineteenth century, particularly in connection with such phenomena as hypnosis, unconscious suggestion, and crowd behavior.[25]

In the same year as Tarde's *Laws of Imitation*, the American psychologist William James published his classic two-volume text *Principles of Psychology*.[26] In this work, James classifies imitation as an instinctual behavior operating early in infancy and essential to what is definitively human. While imitation was not a primary focus of his work, his brief treatment of the subject is nonetheless worth mentioning given his founding influence in the field of psychology. James begins by noting that imitation of facial gestures may begin in childhood as early as 15 weeks. He then elaborates:

> And from this time onward man is essentially *the* imitative animal. His whole civilization depends on this trait, which his strong tendencies to rivalry, jealousy, and acquisitiveness reinforce. '*Nil humani a me alienum puto*,' is the motto of each individual of the species; and makes him, whenever another individual shows a power or superiority of any kind, restless until he can exhibit it himself. But apart from this kind of imitation, of which the psychological roots are complex, there is the more direct propensity to speak and walk and behave like others, usually without any conscious intention of so doing. And there is the imitative tendency which shows itself in large masses of men, and produces panics, and orgies, and frenzies of violence, and which only the rarest individuals can actively withstand.[27]

The nature and importance that James ascribes to imitation is remarkable. However, despite his emphatic treatment of the subject, he did not spend much time developing his initial insights, especially concerning the relationship between imitation, rivalry, jealousy, and acquisitiveness. Still, his more influential theory of ideomotor action asserted that "thinking was for doing," an expansion of Hermann Lotze's view that simply thinking about or perceiving an action automatically made one more likely to do the same thing.[28] Yet curiously, James left out the word "imitation" in his explanation of this principle.

By the end of the nineteenth century, for reasons that are not altogether clear, imitation as a central concept was substantially marginalized by the most influential psychological theories as they began to emerge in the twentieth century. This profound neglect is exemplified by the absence of imitation in the development of Freud's theory of psychoanalysis. Certainly the magnitude and influence of Freud's thought in the twentieth century cannot be overstated, both in the realm of the social sciences and popular culture. It is therefore significant that Freud rarely mentions the function of imitation directly in his many papers and books. From a developmental perspective, "There is no place in Freud's theory of early infancy for imitative self-other reciprocity (primary intersubjectivity)."[29] While his initial theory and treatment approach relied heavily on social influences and incorporated the function of suggestion and hypnosis—developments that were influenced by imitation theorists such as Tarde at the turn of the century—he ultimately abandoned this work in favor of his libidinal theory. In doing so, Freud adhered to an autonomous view of the self that was confined to biological drives and instinctual formulations—a view that he staunchly held until the end of his life. His many followers similarly rejected imitation as a fundamental capacity central to human development and psychological organization.

Like psychoanalysis, learning theories that predominated in the twentieth century—from Skinner's behaviorism to Piaget's cognitive theories of development—assumed that imitation was *not* an innate ability present immediately at birth and therefore did not play a foundational role in the elaboration of more complex cognitive functioning. Early on, Paul Guillaume's definition implied that imitation required at least an elemental level of formal representation.[30] This misconception directed the next generation of cognitive psychologists to uncritically accept imitation as a learned ability emerging later in human development. Wallon, for example, claimed that imitation did not occur prior to 18 months.[31] Thus, the standard view of imitation was that it developed as a result of representational thinking or classical forms of conditioning based on reinforcement.

Piaget, however, did not fully embrace Guillaume's definition, despite the fact that his research was conducted during the same period. Instead, Piaget viewed imitation as "an action by which a model is reproduced, whether the process depends on perception or representation."[32] Nonetheless, like previous theorists, Piaget argued that imitation was a developmental milestone achieved after the first year of life.[33] Piaget's research on imitation, as well as his model of human development and cognitive functioning, became one of the most respected and influential model for almost fifty years.[34]

In conclusion, the inheritance of ancient Greek philosophical conceptions of mimesis that associated imitation primarily with representation, combined with Enlightenment assumptions about an autonomous self, led most modern philosophers and psychologists to ignore or marginalize the role of imitation as a generative and foundational mechanism present at the beginning of life, and which continues to function significantly throughout adulthood. As a result of these influences, many false beliefs about imitation persisted in the cognitive and social sciences. These limiting assumptions had the enduring effect of steering researchers away from imitation as a rich and viable area of investigation on the grounds that they had already understood the phenomenon completely. Such biases, which continued throughout most of the twentieth century, had not been questioned critically until recently, producing disciplines that severely underutilized this most basic aspect of human cognition and intersubjectivity.

THE MIMETIC THEORY OF RENÉ GIRARD

René Girard is a French historian, literary analyst, and cultural theorist. The development of his *mimetic theory*, beginning in the early 1960s, was revolutionary in part for his use of imitation as a unifying concept, despite its neglect among the predominant scientific and interpretive paradigms of the time. Early on, Girard discovered a fundamental relationship between imitation and the unique nonrepresentational qualities of human desire and intersubjectivity. From this foundation, he was able to describe a number of emergent and generative processes that account for many aspects of human life, including the enigmatic nature of human violence and its relationship to the origin and structure of human culture and religion. Not only did Girard recognize such a primordial role of imitation prior to contemporary experimental research, but he did so through investigations into European literature, historical and anthropological writings, and by turning ultimately to religious myths and texts for further evidence of imitative phenomena. His treatment of human imitation is therefore both substantially and methodologically unique.

According to Chris Fleming, Girard has not only produced his own research, but a "research *programme*", one that has "exerted considerable and ongoing influence on a wide variety of work in the humanities and social sciences."[35] There is now a substantial and ever-growing body of secondary literature by scholars who have made use of Girard's ideas on imitation in such fields as anthropology, economics, literary analysis, philosophy, psychology, and theology, to name just a few. As a result, any brief summary will fail to demonstrate the vast interpretive scope and application of Girard's mimetic theory. Nonetheless, a brief survey will be provided that covers some of the major ideas used throughout this book. For a more thorough introduction, the reader is referred to the first part of the interview with Girard at the end of this volume, as well as a number of recent publications that provide engaging and comprehensive overviews of his life and work.[36]

Mimetic Desire, Rivalry, and Violence

In his first book, *Deceit, Desire, and the Novel: Self and Other in Literary Structure*, Girard put forward his first major hypothesis that human desire is mimetic.[37] His initial research centered on the advent of the novel as a new literary genre in nineteenth- and early twentieth-century Europe, and included many of the principal authors, such as Cervantes, Flaubert, Stendhal, Dostoevsky, and Proust.[38] From his comparative analysis, Girard observed that regardless of the author or period, human desire was almost always depicted as a "triangular" phenomenon involving a subject, a model, and a desired object. In contrast to Romantic or Enlightenment conceptions, Girard concluded that these great novelists intuitively understood that human desire was essentially imitated from, or *mediated* by, the desire of others.

Mimetic desire, therefore, refers to what Girard considers the nature of desire itself; that is, human desire is constitutively imitative rather than being an autonomous expression of the individual self. Psychologically speaking, this aspect of his work is a theory of human motivation based on imitation. While biological "appetites" are instinctual and require that animals seek basic elements of survival (e.g., food, sex, shelter, etc.), for humans, the specific form that each element takes is mediated by culture. In other words, our particular desires are determined by what is found desirable by others in our social world (e.g., a popular brand of jeans, a new technological gadget, a social trait admired in a person, a position of power or status, etc.). This is true whether our models exist within our own immediate social sphere (what Girard calls *internal mediation*) or are more distant, where direct social contact may not be possible—such as a celebrity on TV or a religious or political leader who lived hundreds of years ago (what Girard calls *external mediation*).

In either case, only humans look to others in order to learn what is desirable and who to become.

While our great capacity for learning based on imitation has been readily acknowledged for thousands of years, Girard emphasizes that both Plato and Aristotle significantly misrepresented the function of imitation by limiting their definitions to acts of *"representation*—types of behaviors, manners, individual or collective habit, as well as words, phrases, and ways of speaking."[39] So influential were these early formulations that the word "imitation" has remained almost exclusively synonymous with the type of direct copying of behavior or language seen in childhood, or in the mimetic tradition of the arts.[40] In effect, imitation remained "so tied to its origins in the works of Plato and Aristotle that few thinkers before the twentieth century sought to redefine or rethink it in any substantial way."[41]

According to Girard, more fundamental than these largely conscious and representational forms of imitation is our pervasive and nonconscious tendency to imitate the underlying intentions and desires of others. In other words, for Girard, imitation operates at a level that precedes representational thought and affects not only behaviors but also our very felt sense of psychological aliveness and motivation. What this means is, we usually do not recognize the ongoing function of imitation on our affective experience of desire. Instead, what we usually experience is a linear, or autonomous, process by which we are suddenly motivated or curious about an object. The essential misrecognition in this process is that it is a *model's desire* that effectively creates interest and value in a particular object, and not the object itself. From this perspective, human desire is not innate or "romantic" in the sense of originating spontaneously, or "springing up" from within oneself—as in Descartes's notion of the autonomous self, or Freud's instinctual-drive theory. Instead, our unique and pervasive capacity for imitation effectively creates such an illusion.[42]

Furthermore, Girard perceived that mimetic desire had a generative effect on human motivation; that is, mimesis is mutually reinforcing, whereby one person's desire compounds the other's desire, creating an increasingly concentrated interest in an object and varied attempts at possessing it. This bidirectionality, or *double mediation*, inherent in reciprocal relations has the capacity to escalate into intense passion for potentially any object, whether or not it has any "real" or inherent value apart from its significance in the social sphere.[43] This realization convinced Girard that the mimetic nature of desire had many broad and far-reaching consequences in all aspects of human life, particularly in contributing to human competition and conflict.

Once again, Girard regards Plato's work on mimesis as ultimately insufficient for understanding the more dynamic aspects of human imitation. "What is missing in Plato's account of imitation is any reference to kinds of behavior involved in appropriation [the process of gaining possession of something exclusively for oneself]. Now it is obvious that appropriation figures formidably in the behavior of human beings, as it does in that of all living beings, and that such behavior can be copied. . . . It was Plato who determined once and for all the cultural meaning of imitation, but this meaning is truncated, torn from the essential dimension of acquisitive behavior, which is also the dimension of conflict."[44] Indeed, modern theories of imitation, beginning with Tarde, have primarily emphasized the socially cohesive aspects of imitation, including its essentially positive role for learning and cultural innovation, all the while remaining unaware of, or not adequately exploring (as was the case with William James), its essential relationship to human acquisitiveness, envy, competition, and conflict.[45]

For his part, Girard has articulated how the mimetic process between two persons may progress through three identifiable stages from mimetic desire, to mimetic rivalry, and ultimately to mimetic violence.[46] Beginning with mimetic desire, which creates desirability for something based on the influence of a model (whether it is a tangible possession or something more abstract, such as social status), *mimetic rivalry* results from the attempt to acquire and appropriate the same object. According to Girard, because only one person can ultimately appropriate or possess an object exclusively for oneself, *acquisitive* or *appropriative mimesis* regularly leads to conflict. As a result, participants become *model-obstacles*; that is, each is simultaneously both a model for the other's desire as well as the primary obstacle to fulfilling it, effectively generating a *double-bind* situation. Girard calls this stage "mimetic rivalry" because of a transformation that takes place in the focus of each participant's desire: the fundamental shift is from the object of desire to the opposing rival. In other words, we become so caught up in the rivalry itself that we often forget what it was we were fighting over in the first place, since defeating one's rival has become the primary desire.

As this conflict progresses and intensifies, individuals become increasingly undifferentiated with respect to their movements, utterances, accusations, and sense of identities. They become mirrors of one another, or what Girard calls *monstrous doubles*, referring to the ultimate convergence of exact reciprocity in the midst of intense conflict. However, this is not the subjective experience of each person, since in the moment each individual is absolutely convinced that they are completely different from the other, while in fact both individual behaviors and intentions may be exactly alike.

This fascination/abhorrence vortex of emotional arousal is one of the most powerful of all human experiences, and one of the most difficult to stop or back away from once it has crossed a certain threshold of intensity. As a result, *mimetic violence* is the last stage in the deepening crisis of mimetic conflict described by Girard. As individuals become more desperate in their attempts to differentiate themselves from their rival, their behaviors eventually erupt into physical violence. This violence in turn is similarly imitated and increased reciprocally with each act. Ultimately, mimetic violence can become so intense that the only perceivable resolution for establishing differences is to kill one's rival. Indeed, there is no greater distinction than between one living and one dead.

In his first book, Girard concluded that the relational drama and conflict central to the great novels he was studying reflected the *novelistic truth* of the human condition; that is, without a way to come to terms with our mimetic dependency on one another, our relationships will forever perpetuate the *romantic lie* of self-autonomy—a lie that can only be maintained through infinite forms of deceit, hypocrisy, greed, isolation, and even death.

The Scapegoat Mechanism and the Origins of Human Culture and Religion

After formulating his literary theory of mimetic desire in his first book, Girard turned to anthropology in order to understand the nature and presence of imitation and violence in early cultures, especially in archaic religion and Greek tragedy. His findings were first published in 1972 in *Violence and the Sacred* and then elaborated on further in his subsequent book, *Things Hidden since the Foundation of the World*.[47] In *Violence and the Sacred*, Girard put forward his second major hypothesis, the *scapegoat mechanism*, which was an analysis of the function of archaic rituals (including human sacrifice), prohibitions, and myths in the genesis and maintenance of human culture and religion.

Anthropologically speaking, this aspect of Girard's work is a theory of human origins. He wanted to understand how human culture was able to come into being and ultimately survive and flourish, given the evolving and immensely problematic forms of imitation. Because violence is one of the most imitative of all human behaviors, once initiated it has the potential of spreading quickly to those within its proximity, upsetting existing bonds of community, and leading to cyclical acts of revenge and violent group contagion. In the absence of any instinctual elements of restraint, such as dominance patterns found in other animal societies, Girard reasoned that the process of hominization—the evolutionary shift from nonhuman primates to humans—could not have taken place unless there was a way to contain and regulate mimetic rivalry. According to Girard, social contracts would have been impossible given the prelinguistic nature of purely mimetic relations at this stage of cultural evolution. Without some "natural" way to limit mimetic violence and bring it to an end, the likely result would have been the extinction of our emerging species.

For Girard, the natural or "spontaneous" phenomenon that made social stability possible, and ultimately a cohesive human culture, was the mimetic displacement of violent tensions among members of the community onto a *surrogate victim.*[48] While complex and wide-ranging, Girard's anthropology asserts that archaic culture and religion have their origin in the same type of event: a *mimetic crisis of undifferentiation* (all against all) that polarized into a collective murder (all against one). This event had the effect of creating an unprecedented sense of communal bonding that eliminated group tensions while at the same time establishing important social distinctions: that between a community and its antagonist. This, according to Girard, is the origin of human sacrifice. In the attempt to prevent such spontaneous episodes of uncontrollable violence, this *scapegoat mechanism* and its resolution was repeated and thus made available for imitation and cultural transmission on a ritual basis.

Girard's explanation of the origin and function of sacrificial rituals— arguably the earliest and most universal of all human institutions in archaic societies—accounts for the paradoxical relationship between violence and religion that is still largely misunderstood today:

> If my analysis is sound, far from being the cause of our violence, archaic religions are, or rather were, first a consequence of that violence and, secondly, our primary protection against it. During the longest part of our history or pre-history, they enabled human communities to survive their own violence. Archaic religions are essentially combinations of prohibitions and sacrifices. Prohibitions forbade violence directly, but they often failed and, when they did, archaic communities fell back upon their second line of defense, sacrifice. The paradox of archaic religion is that, in order to prevent violence, it resorted to substitute violence.[49]

As a result of this understanding, Girard has analyzed a great number of archaic myths, particularly myths of origin, in order to provide further support for his hypothesis. Girard maintains that once the capacity for language had evolved, the role of such myths was to both conceal and reveal certain aspects of the culture's founding violence. According to Girard, scapegoating rituals only work if the victim is perceived as guilty. Mythological texts, therefore, never present an explicit theme of scapegoating; instead, myth camouflages scapegoating while representing patterns of meaning in stories of gods, ancient heroes, foundations of social order, rituals, etc.

Even if the events of any myth may be unbelievable to modern critical thinking, Girard argues that they do have empirical and historical referents at the generative level, including structural signs of scapegoating, or what he has termed *stereotypes of persecution*.[50] These structural elements include a social crisis involving the loss of distinctions felt to be necessary for social order; the identification of a particular individual (animals, beasts, gods, etc.) onto whom the alleged crimes undermining law and order are transferred; some sign or distinction that sets the victim apart from the group, such as foreign status, a physical abnormality, or some other sign of marginality; the expulsion or murder of the individual; and the restoration of social order and harmony and the creation of the "world" or culture. From this perspective, myths are a later cultural development that allowed archaic communities to have a narrative that accounted for the origin and continued practice of its sacrificial rituals that at the same time did not undermine their perpetuating mechanism: collective violence directed toward arbitrary and substitute victims.[51]

The Revelation and Reversal of the Scapegoat Mechanism in Human History

In addition to postulating human origins in scapegoating violence, Girard has also provided a framework for understanding the centrality of this mechanism in the unfolding events of human history, including the modern world's awareness of scapegoating violence and its unprecedented sensitivity to social victims. In addition to his analysis of myths of origin in archaic culture, Girard has also analyzed a wide range of religious, anthropological, and philosophical texts—including Greek myths and tragedies, the Hebrew Bible and the Christian Gospels, texts of persecution in the Middle Ages, and modern philosophy—in order to trace humanity's growing awareness of, and attempts to do away with, scapegoating as a viable basis for social organization.[52] These additional historical and theological considerations represent the last major development of Girard's mimetic theory.

Girard's analysis of the Hebrew Bible and Christian Gospels is perhaps the most controversial aspect of his work. While applying his mimetic insights to a number of religious and cultural texts, Girard discovered that the Hebrew

Bible and Christian Gospels represented a unique and revelatory movement away from sacrificial forms of cultural maintenance.[53] Girard claims that while these texts include all of the structural elements of archaic myth, including collective violence and sacrificial rituals, they also feature religious leaders and communities who were openly troubled by these events and began to view their history from the standpoint of the social victims rather than the persecuting, or sacred social order. In essence, for Girard, both the Hebrew Bible and Gospel narratives chronicle the evolution of humanity from our origins in sacrificial rituals to our modern struggle to empathize and identify with social victims and thereby come to terms with our own violence.[54]

In summary, Girard's theories of mimetic desire and the scapegoat mechanism attempt to illustrate how in the process of hominization human mimetic behavior was inherently unable to establish relational order apart from the use of violence. Archaic religion and culture are seen as sharing the same origin and are intimately related in their service to acts of scapegoating upon which human relational stability was forged and maintained on a ritual basis. Human history similarly reflects our ongoing struggle to come to terms with our own imitative nature, which is the source of our great capacity for learning and cultural advancement as well as, paradoxically, the persistent cycles of violence that continue to underlie our greatest social, political, and religious concerns.

THE NEW SCIENCE OF HUMAN IMITATION

At the time Girard articulated the major tenets of his mimetic theory, there was a considerable gap between his understanding of the primordial role of imitation in human life and the level of interest and research within the cognitive and behavioral sciences. In fact, "until the 1970s the term 'imitation' did not even appear as a keyword in reference bases such as Psychological Abstracts," and "the existence of immediate imitation in development was hardly suspected and its role was ignored."[55]

Since the late 1970s, however, there has been a dramatic surge of interest in imitation across a wide range of disciplines, producing unprecedented support and clarification of the elemental role of imitation in human life. In a similar way that Girard's mimetic theory overcame many previous limitations in our conceptualization of imitation, so has contemporary experimental science discovered the generative function of imitation and redefined its role in human cognition and culture, from both developmental and evolutionary perspectives. Because this body of research is voluminous—spanning over 30 years of empirical work—what follows is a brief presentation of some of the major findings and conclusions from developmental psychology, comparative psychology, neurophysiology, cognitive neuroscience, and social psychology.

Developmental Psychology—Neonatal and Infant Imitation

Developmental psychology played a pioneering role in changing the depth and scope of recent imitation research. This is perhaps best exemplified by the seminal work of Andrew Meltzoff and Keith Moore.[56] While testing Piaget's developmental stages of infant preverbal learning, Meltzoff and Moore discovered that newborn infants were able to learn via imitation immediately at birth, including the reproduction of facial expressions that were not based on conditioning or the triggering of innate responses.[57] This finding essentially debunked what was thought to be a disconnect in infancy between perception and action, and self and other, as espoused by such dominant theorists as Freud and Piaget. In addition, because "the mechanism by which infants can connect the felt but unseen movements of the self with the seen but unfelt movements of the other" were largely unknown at the time, the discovery of immediate imitation in infancy opened the door to a new and exciting field of inquiry concerning the starting state of human cognition, including the nature and role of imitation in human development.[58]

Imitation used by preverbal children has since been studied extensively and has shown to be the basis for primary forms of human intersubjectivity, including social and affective coordination,[59] nonverbal communication,[60] and self-other identity and differentiation.[61] Early imitative exchanges allow children to communicate intentions, negotiate turn-taking and role-switching, share in pretend play, and collaborate in joint projects.[62] In essence, the contingency and reciprocity of intersubjectivity afforded by imitation in early infancy is understood as the basis for the emergence of more complex social and interpersonal skills.

In addition to immediate imitation in infancy, the operation of "deferred imitation" has also been investigated.[63] Deferred imitation refers to the delayed re-presenting of past novel events, an ability long believed to be a developmental milestone occurring at 18 months of age.[64] More contemporary research has revealed, however, that as early as six weeks, infants can immediately encode a novel act or gesture from a brief single exposure and, when prevented from imitating it immediately, can reproduce it after delays as long as 24 hours.[65] At 12 months of age, children can successfully defer imitation after a delay of four weeks, and by two years of age the delay can be four months or longer.[66] Together, immediate and deferred imitations are considered powerful and advantageous learning tools for humans because they avoid time-consuming trial-and-error learning. As a result, imitation allows the child to adapt to novel situations and produce increasingly complex behaviors, including the appropriate use of language and cultural skills.[67]

Essential to these social intelligences is our ability to infer and think about the mental states of others, including the goals, intentions, and desires that underlie their observable actions. Understanding that others have minds like our own, including our ability to "read" the minds of others, is referred to in the literature as "theory of mind" (ToM).[68] Since imitation has been found to operate powerfully prior to language and representational thinking, imitation researchers have become increasingly involved in the debate concerning the origin and mechanisms of ToM in humans. According to Colwyn Trevarthen and his colleagues:

> Recent speculations about how a "theory of mind" develops in children constitute a real advance by recognizing that what goes on in minds is naturally of interest to humans. But, these models have not, we believe, much clarified the problems of how sympathetic awareness begins. They merely rephrase the verbal representational hypothesis in mentalistic or cognitive science (machine intelligence) language. The theory of mind debate is leading to clarification of important steps in the development of human intersubjectivity after language has been mastered. However, the basic ability to imitate remains to be understood. It is independent of both linguistic and rational representations, and it is not a symbolic formulation of machine "thinking." Mimesis generates symbols, not the other way around. Imitation is part of the needed explanation.[69]

The idea that imitation, or mimesis, plays a primary role in the genesis of mental representation, rather than the other way around, signifies a dramatic shift in perspective from previous cognitive and philosophical views.[70]

Comparative Psychology—Imitation in Animals

The study of social learning in animals is another major area of research that has contributed significantly to a more contemporary understanding of the unique imitative abilities found in humans, including their role in the evolution of human cognition and culture.[71] The above studies on immediate imitation in human infants inspired similar inquiries among comparative psychologists. Using a variety of innovative methodologies and theoretical applications, copying behaviors in animals, especially nonhuman primates, have since been researched and compared with those found in human infants and adults.

Contrary to popular belief ("monkey see, monkey do"), the comparative research has shown that "true imitation" in nonhuman primates is difficult to find, and that additional copying behaviors are not nearly as complex or efficient as those found in humans.[72] There is considerable debate, however, centering on definitions of imitation, which can be highly varied and specific, as well as methodologies for eliciting imitative behavior in nonhuman primates. For example, it is questionable as to whether or not copying behaviors found in monkeys who were raised immersed in human cultural settings, as well as those that had human subjects as imitative models, accurately represent the cognitive capacities of nonhuman primates alone. While it is clear that there are evolutionary precursors to imitation in other primates, many studies that claim to demonstrate imitation in nonhumans have since been explained in terms of more basic processes, such as response facilitation.[73]

In general, the human capacity for imitation is considered qualitatively and quantitatively distinct in comparison to other animals—so much so, that Richard Byrne, one of the leading researchers in the study of primate cognition, equates our advanced capacity for imitation with magic:

> That element of magic . . . explains the academic popularity of imitation today—in the face of a persisting lay view that imitation is a mere sham of intelligent behavior, a cheap trick. The magic is two-fold. First, how can the imitator recognize that an action it performs is "the same" as that observed in another? The perspectives may be very different, with little visual or auditory similarity; it is magic. Second, how can a package of skills be "transferred" from one individual's repertoire to that of another? If this trick can be done, it offers a rich method for the transmission of useful procedures between individuals, and even across generations. But how might a complex of goals, rules, processes, and schemata be obtained from simply seeing an act done by another? It is magic.[74]

This renewed view of imitation as a highly complex and seemingly "magical" ability has stimulated theoretical developments over the last several decades that seek to provide a more comprehensive account of the continuity and divergence of cognitive skills across primate species, including the role of imitation in the evolution of the unique features of the human mind.[75]

Emerging from such analyses are several general theories of the origins of human culture that recognize the capacity for imitation, or mimesis, as a defining characteristic.[76] Merlin Donald, for example, has proposed three distinctive stages, or transitions, of human cultural evolution beginning with "mimetic culture" (defined by action metaphor), to "mythic culture" (defined by language and symbolic representation), and finally to "metaphoric culture" (defined by an external symbolic universe). According to Donald, mimetic culture is the first definitively human stage that separates our cognitive and social skills from those of other primates, including chimpanzees, who function within a more limited "episodic culture" that is confined to event perceptions. Mimetic culture, on the other hand, includes the collective reenactment of events, gestures, and skills, all of which precede the development of language, symbolic functioning, and more personal forms of reflective or private thought. Donald explains:

> Mimesis must have come early in hominid prehistory because it was a necessary preadaptation for the later evolution of language. It provided the underpinnings of social connectivity and conventionality. It took the primate mind one step further in the direction of improved social coordination and collective cognition. The group was primary, and thus having an accurate sensitivity to group feelings was a survival-related skill. Mimesis is still the elemental expressive force that binds us together into closely knit tribal groups. Of all our human domains, mimesis is the closest to our cultural zero point.[77]

Neurophysiology—Mirror Neurons

Perhaps more than any other recent discovery, research on mirror neurons has generated a surge of interest in imitative phenomena, both from scientists and lay audiences alike.[78] The Italian team of neurophysiologists led by Giacomo Rizzolatti first reported on mirror neurons in the mid-1990s from their research with macaque monkeys.[79] Using single-cell electrodes to record neural activity, they found that individual neurons that were activated while the monkey was performing a particular goal-directed motor sequence, such as grasping a piece of food with its hand, would equally become activated while the monkey simply observed the experimenter performing the same action. The responsiveness of these neurons to both direct motor movement as well as purely visual information of the same motor sequence suggested a direct resonance or common coding of observation and execution between participant and observer.

This discovery has been considered revolutionary in part for its contribution to understanding many unanswered questions concerning primate social cognition, including the neurophysiological basis of imitation in humans. Rizzolatti and his colleagues explain why: "The novelty of these findings is the fact that, for the first time, a neural mechanism that allows a direct matching between the visual description of an action and its execution has been identified. Such a matching system constitutes a parsimonious solution to the problem of translating the results of the visual analysis of an observed action . . . into an account that the individual is able to understand."[80] As a result, mirror neurons provide support for earlier models of immediate imitation proposed by developmental psychologists, based on behavioral observation. They also take our understanding of mimetic phenomena to a whole new level of functionality and research—that is, the subpersonal level of cortical organization and neural integration.

In addition to the neural correlates of imitation, the functional significance of mirror neurons is also thought to contribute to many interrelated facets of social cognition in humans, including empathy, affective resonance, action representation, communication and language, and theory of mind.[81] Additional theoretical speculations stemming from research on mirror neurons address their significance in helping to explain not only the underlying mechanisms of such skills but also their evolution across species. For example, it has been argued that differences between human and nonhuman primates may be due more to cortical "rewiring" rather than simply brain size or the acquisition of unique brain structures.[82] In other words, mirror neurons and the evolution of a more complex "mirror system," or imitative brain, may have contributed significantly to a wide-scale cerebral reorganization, allowing for the coevolution of more complex social and representational skills across primate species.[83]

Cognitive Neuroscience—Brain Imaging

In the 1990s, the innovative methodologies of cognitive neuroscience emerged as technological advances allowed researchers to study more detailed activity of the live human brain.[84] During this period, the discovery of mirror neurons prompted a number of cognitive neuroscientists to utilize brain-imaging technology to further understand the mechanisms of behavioral imitation, as well as the neurocognitive basis of human intersubjectivity.

Marco Iacoboni and his colleagues have used functional Magnetic Resonance Imaging (fMRI) with human subjects to show that merely seeing an action performed significantly activated the same neural areas *as if* the subjects were performing the same action.[85] Such studies have been used to argue for the presence of a similar mirror neuron system in humans that serves as the neural basis for imitation.[86] However, while mirror neurons may be a neural mechanism underlying our ability to imitate, human imitation is vastly more complex than the in vivo resonance of visual-motor information. For instance, even though monkeys have mirror neurons, they have not been found to be effective imitators. Mirror neurons therefore do not explain the whole of human imitation. As a result, Iacoboni and his colleagues have drawn from a wide range of imaging data to identify what they consider to be the "minimal neural architecture" for imitation in humans. This architecture includes coordinating circuits between the temporal, parietal, and frontal lobes of the brain that provide visual, somatosensory, and goal-oriented information to and from mirror neurons in these corresponding brain regions.[87]

Another prominent line of research using brain-imaging technology has looked at the relationship between imitation, empathic resonance, and self-other differentiation. Research by Jean Decety and colleagues, for example, has demonstrated that similar brain regions are activated when one is simply *imagining* the same action from either perspective (observing or participating), providing further information on the neural substrates of imitation and empathic resonance.[88] Using both fMRI and Positron Emission Tomography (PET), imaging studies have been used to identify specific brain regions that are thought to be involved in shared representational states as well as differentiating between self and other with respect to imitating or being imitated.[89] Decety and Sommerville explain their overall findings:

> Consistent with research and theoretical claims from developmental and social psychology, representations of aspects of the self both overlap with representations of other and are distinct from such representations. Common and distinct representations of self and other extend along many dimensions of self and other processing: from action recognition to mental state understanding. Indeed, such shared representations, including beliefs, unify the cognitive and motivational processes that constitute the contents of culture. These findings shed light on the nature of the self as both special and social, unique and shared.[90]

In other words, when two individuals are involved in reciprocal imitation, it is thought that they share similar representations of the action, as well as code-specific information relevant to their own point of view.

Beyond the imitation of surface behaviors, imaging technology has also been used to study the role of the mirror system with respect to emotions, tactile sensations, and the reading of another person's intentions. Using fMRI, Christian Keysers and colleagues have found that both feeling disgusted and watching someone else look disgusted activated a similar segment of the anterior insula, a particular region of the olfactory area of the brain.[91] They also found similar results for activation of the somatosensory cortex during direct tactile stimulation and simply viewing someone else being touched.[92] Finally, Iacoboni and his colleagues present evidence for the function of the mirror system in reading intentions that underlie surface behaviors.[93]

Social Psychology—Unconscious Imitation

Lastly, research from social psychology and the study of adult social cognition has also played a major role in recent interest and understanding of the significance of human imitation, including its pervasive operation at an unconscious level and throughout the life span.[94] In line with James's principle of ideomotor action developed at the end of the nineteenth century, as well as recent research on the mirror system in humans from cognitive neuroscience, experimental evidence from social psychology demonstrates how just thinking about or perceiving a certain kind of action, even unconsciously, automatically increases the likelihood of engaging in that particular behavior.

According to Dijksterhuis and Bargh, the human default tendency to act in the same way that we see or imagine others acting "flows directly from a fact of mental representation and organization—that perceptual and behavioral representations for the same action overlap."[95] This "perception-behavior expressway" of human imitation is so pervasive that its effects have been found to cut across practically every dimension of human social relations, including speech and verbal mimicry (e.g., words, clauses, grammar, rates of speech, tones of voice, syntax, etc.), facial expressions, behavioral matching (postures, mannerisms), and emotions and mood.[96] Such broad imitative influences have been referred to as the *chameleon effect*[97] and do not depend on any conscious goal, making imitation the "social glue" that unconsciously guides self-regulation and social affiliation throughout the lifespan.[98]

Additional research from social psychology demonstrates that humans are uniquely social animals not simply because of their ability to imitate, but because of "the wide range of behavior and even entire behavioral patterns that they imitate."[99] For example, Dijksterhuis distinguishes between two imitative pathways.[100] The first is considered the "low road" and refers to behaviors that are imitated through direct and literal observation. This is the standard view of seeing and then doing what another does. The second pathway, what Dijksterhuis calls the "high road" to imitation, refers to the imitation of complex patterns of behavior that are mediated unconsciously by the activation of traits or stereotypes. The high road to imitation is considered much more broad and pervasive than the low road and can be elicited simply by being exposed, even unconsciously, to semantic content, such as simple words, associated with that behavior.

Studies using implicit, or unconscious, primes prior to performance on particular tasks have been especially revealing of the high road to imitation. For example, implicit priming with semantic content alone (i.e., words associated with a certain stereotype or trait) has been shown to significantly alter a wide range of simple behaviors and attitudes (walking slower or behaving rudely),[101] more complex goals and behaviors (performing better or worse on a test of general knowledge),[102] and interpersonal goals (performance vs. cooperativeness).[103]

In essence, cognitive and social psychologists consider imitation to be the "default social behavior," so much so that not imitating is the exception in most of our social interactions.[104] In fact, the "shared rhythms of behavior" and psychological states produced by such pervasive and ongoing forms of imitation are thought to be a primary motivation for human interaction in and of itself.[105] However, researchers also note how difficult it is for most people to believe that they are imitating and being influenced by others in such broad and automatic ways, even when they are presented with evidence that indicates otherwise. The tendency to deny our own imitative behavior may be due to its nonconscious operation, combined with the fact that such knowledge threatens the notion of our selves as being primarily conscious and in control of our own actions.[106]

In conclusion, over the last 30 years, experimental researchers from a wide range of disciplines have produced a dramatic array of data elucidating the uniquely generative and foundational qualities of imitation in humans, including its significance throughout the lifespan from both developmental and evolutionary perspectives.

WHERE ARE WE NOW?

It is clear that a new environment exists in which Girard's mimetic theory and experimental research on human imitation can be further developed and explored together. This environment is a multidisciplinary field of study that promises to address gaps in both fields, as well as strengthen existing areas of knowledge through convergent support from a wide range of disciplines.

With psychological mimesis as his foundation, Girard has made a number of bold and influential claims about imitation and its relationship to human desire, intersubjectivity, and the origins of human culture and religion. However, any theory attempting to cover this much ground will undoubtedly draw a sizable amount of criticism, and Girard's work has been no exception. The absence of sufficient empirically based models of imitation at the psychological level alone has allowed many critics to reject outright his mimetic theory as a whole. While Oughourlian, Webb, and Garrels have made reference to several imitation studies in support of Girard's theory, the full extent of potential integration between the two has yet to be explored.[107]

Several authors have previously noted the importance of furthering such an analysis.[108] Webb, for example, has argued that while Girard's work has many broader anthropological implications, the most essential aspect of his theory is that of psychological mimesis: "There is good reason to think not only that the latter deserves a careful hearing but also that it should prove widely useful for psychologists in providing an explanatory framework for the sorts of systematic relationship that many are currently investigating."[109] While Paisley Livingston has attempted a systematic presentation of Girard's ideas on mimesis, concluding that "Girard's hypotheses about the pragmatics of imitative interaction, and more specifically, about the influence of imitation on motivation, represent a unique and significant contribution that merits a careful examination,"[110] such an examination has not taken place through an engagement with research from the experimental sciences. This is remarkable when one considers the enormous output of imitation research in the last decade alone.

The ability of mimetic scholars to effectively address the various disciplines they claim to interpret may therefore depend upon their level of engagement with contemporary imitation research. This relationship is important because of the many gaps that continue to exist in mimetic theory, which have yet to be addressed in such a way as to garner sufficient scientific support for its claims. The most obvious gap is the question of *how* the mechanisms of imitation actually function in the human brain and in the interpersonal matrix where they are found, including their developmental and evolutionary pathways. Admittedly, identifying the neurocognitive mechanism of imitation has never been the goal of Girard's research, nor has it been possible until recently. Rather, Girard has always been more interested in the broader consequences of imitation in human culture and society. Regardless, contemporary empirical research on imitation now "illustrates promising methodologies for interactive collaboration among the cognitive and social sciences and philosophy."[111] As a result, going forward will require mimetic scholars to articulate the types of hypotheses that the mimetic theory provides, while attending to relevant data from the experimental sciences. If this can be done well enough, then Girard's theory may suggest new empirical research, and the empirical research may in turn suggest transformations of Girard's mimetic theory.

On the other hand, while experimental research provides an abundance of unprecedented support for, and clarification of, the foundational role of psychological mimesis in human life, this body of work has yet to be used—and made substantially more innovative—by exploiting the many obvious links with Girard's mimetic theory. The work of Girard and other mimetic scholars provides empirical researchers with an already elaborated explanatory framework for thinking about the ways in which imitation is foundational to a wide range of human phenomena, including its generative effects on human desire, competition, conflict, and violence, as well as the overlooked contribution that such mimetic effects have on a wide range of social, cultural, and religious domains.

In contrast to such broad concerns found in the work of mimetic scholars, imitation researchers have primarily studied imitation within two-person interactions in order to understand its functional architecture and developmental pathways as well as cognitive links between human and nonhuman primates. This work has typically involved dyadic, short-lived, and often unidirectional imitative interactions between conspecifics.[112] Empirical researchers have therefore only begun to explore mutually influencing and reciprocating feedback loops between persons and over longer periods of time, including the role of imitation on motivation throughout the lifespan, as well as many broader social, cultural, and religious implications.

For example, the following questions indicate frontier areas of research considered by imitation researchers, which overlap significantly with those already explored in depth by mimetic scholars:

> Imitation is a critical locus for understanding the ecology of human cognition and norms: the dynamic interactions between cognitive processes and sociocultural processes. . . . Are innate or cultural deficits primarily responsible for . . . violent aggression? Can individual responsibility itself be understood, compatibly with an innate human tendency to imitate, in partly ecological terms? . . . How should we respond to the irony of imitation: that the capacity for imitation appears to be a distinctive feature of human nature and may well be part of the basis for other distinctive features

of human nature, such as mind reading and language, which together set us apart from other animals? Yet at the same time our innate, automatic tendencies to imitate can also threaten our conception of ourselves as autonomous and deliberative in ways that no other animals are.[113]

Broadly speaking, the whole of Girard's mimetic theory is an elaboration of that very irony; namely, that our imitative nature not only affords us those uniquely human forms of social bonding, identification, and intelligence, but it also threatens those affordances at the same time, making human culture equally susceptible to its own unique forms of relational conflict, misunderstanding, and violence—violence that in turn generates its own forms of social bonding and identity. In this respect, and many others, the empirical literature appears unaware of key aspects of Girard's theory.

Perhaps the most important insight of Girard's that deserves to be considered is the role that reciprocal mimesis has in generating acts of social competition, rivalry, and violence. As far back as 1979, Girard offered the following critique: "If you survey the literature on imitation, you will quickly discover that acquisition and appropriation are never included among the modes of behavior that are likely to be imitated. If acquisition and appropriation were included, imitation as a social phenomenon would turn out to be more problematic than it appears, and above all conflictual."[114] Indeed, a contemporary survey will reveal the same results: that this conflictual aspect of human imitation has not been addressed by the cognitive or social sciences. If a connection is made between imitation and violence, it is typically done so in relation to "copycat" behaviors, either through social modeling or violence portrayed in films and the media.[115] Despite the fact that this phenomenon is important and deserves to be addressed, the connection made between imitation and violence overlooks how our imitative nature facilitates initial acts of violence in the first place, before there is any violence to imitate.

From the perspective of Girard's mimetic theory, human violence is an emergent phenomenon based on our uniquely evolved capacity for imitation. If this is true, how did early humans survive their growing propensity for intense social competition, rivalry, and violence? Humans have not only survived their own violent tendencies, but have flourished with ever increasing cultural complexity and technological innovations. Yet at the same time, the threat of violence now seems ever present—violence of an unprecedented magnitude. So how is it that we continue to manage these highly mimetic potentials? What factors enable individuals and communities to shift from what would be mutually assured destruction to transformational acts of positive reciprocity? These and other questions deserve more detailed exploration from a wide range of perspectives.

Perhaps the greatest obstacle that has prevented Girard's work from reaching a wider audience, particularly in the United States, and especially within the empirical sciences, is his argument for a "universal anthropological theory, combined with the position that the deepest insights of Western Culture stem from Biblical revelation."[116] His position that religious traditions in general offer invaluable insights into our human nature has produced no small amount of suspicion in academic circles that ignore or harbor contempt for such religious influences. Girard summarizes his own view on this issue in the following manner:

> Even if my observations are too sketchy to convince you that the mimetic theory of religion is the breakthrough I believe it is, you will agree, I hope, that even the most obviously "untrue" religions are worthy of our respect. Archaic religions are not simply false explications of the universe. They always had more urgent business to attend to than satisfying the curiosity of idle men. They have always been in charge of keeping the peace. Even if they had to resort to violent means to reach their goal, these means were not really their own invention; they were provided more or less ready-made by the spontaneous course of human relations. We cannot condemn these religions as something alien to our modern humanity. Even as we try to do better than the old religions did, we understand that the task is infinitely more difficult than it was thought a hundred years ago. The violence we would love to transfer to religion is really our own, and we must confront it directly. To turn religions into the scapegoats of our own violence can only backfire in the end.[117]

Indeed, the role of religion in society, and the human capacity for both immensely altruistic, as well as terribly violent, acts of social behavior are two of the most significant and pressing topics of our contemporary world. How we understand the two, including their relationship to one another, has enormous bearing on the future survival of humanity and the world in which we live. For this reason, we find it particularly important that Girard's ideas on human imitation, violence, and religion be critically readdressed in light of new evidence from contemporary empirical research, and within an academic community that is open to understanding religion as a potentially positive and important factor in human flourishing.

ACKNOWLEDGMENTS

I would like to thank both Mark Anspach and Andrew Meltzoff for their review and helpful comments of the completed manuscript of this chapter..

NOTES

1. Chris Fleming, *René Girard: Violence and Mimesis* (Cambridge, UK: Polity Press, 2004), 1.

2. For a more comprehensive history of the concept of imitation, or mimesis, in Western philosophy, see Matthew Potolsky, *Mimesis* (New York: Routledge, 2006); Arne Melberg, *Theories of Mimesis* (Cambridge: Cambridge University Press, 1995); and Gunter Gebauer and Christopher Wulf, *Mimesis: Culture—Art—Society*, trans. Don Reneau (Los Angeles: University of California Press, 1995).

3. James S. Ackerman, *Origins, Imitations, Conventions: Representation in the Visual Arts* (Cambridge, MA: MIT Press, 2002).

4. John D. Boyd, S.J., *The Function of Mimesis and Its Decline* (Cambridge, MA: Harvard University Press, 1968), 8.

5. Jacqueline Nadel and George Butterworth, "Immediate Imitation Rehabilitated at Last," in *Imitation in Infancy*, ed. Nadel and Butterworth, 1–5 (Cambridge: Cambridge University Press, 1999), 2.

6. Ibid.

7. Richard McKeon, ed., *The Basic Works of Aristotle* (New York: Random House, 1941), 1457.

8. Boyd, *The Function of Mimesis*, 19.

9. Ackerman, *Origins, Imitations, Conventions*, 129.

10. Karl F. Morrison, *The Mimetic Tradition of Reform in the West* (Princeton, NJ: Princeton University Press, 1982).

11. Thomas à Kempis, *The Imitation of Christ*, trans. Harold J. Chadwick (North Brunswick, NJ: Bridge-Logos Publishers, 1999).

12. Ackerman, *Origins, Imitations, Conventions*, 136.

13. Ibid., 132.

14. Boyd, *The Function of Mimesis*, 8.

15. Stephen Gardner, *Myths of Freedom: Equality, Modern Thought, and Philosophical Radicalism* (Westport, CT: Greenwood Press, 1998), 53.

16. Gardner, *Myths of Freedom*, 60.

17. Antonio Damasio, *Descartes' Error: Emotion, Reason, and the Human Brain* (New York: Harper Perennial, 1994).

18. Andrew Meltzoff and Wolfgang Prinz, "An Introduction to the Imitative Mind and Brain," in *The Imitative Mind: Development, Evolution and Brain Bases*, ed. Meltzoff and Prinz (Cambridge: Cambridge University Press, 2002), 1–15.

19. Potolsky, *Mimesis*.

20. Gardner, *Myths of Freedom*; Jeffrey Stout, *Ethics after Babel* (Boston: Beacon Press, 1988); Stephen Toulman, *Cosmopolis* (New York: Free Press, 1990); P. J. Watson, "Girard and Integration: Desire, Violence, and the Mimesis of Christ as Foundation for Postmodernity," *Journal of Psychology and Theology* 26, no. 4 (1998): 311–21.

21. See Vittorio Gallese, "The Two Sides of Mimesis: Mimetic Theory, Embodied Simulation, and Social Identification," and Jean-Pierre Dupuy, "Naturalizing Mimetic Theory," both in this volume, for a discussion of philosophy in the modern era.

22. See Jean-Pierre Dupuy, "Naturalizing Mimetic Theory," in this volume concerning the impact of the Enlightenment tradition.

23. Gabriel Tarde, *The Laws of Imitation*, trans. Elsie Worthington Clews Parsons (New York: Henry Holt and Company, 1903), xiv.

24. Ibid., 77.

25. See the work of James M. Baldwin, "Imitation: A Chapter in the Natural History of Consciousness," *Mind* 3 (1894): 26–55; Baldwin, *Development and Evolution* (London: Macmillan, 1902). See also Eugene Webb, *The Self Between: From Freud to the New Social Psychology of France* (Seattle: University of Washington Press, 1993).

26. William James, *The Principles of Psychology*, vols. 1 and 2 (Henry Holt & Co., 1890).

27. James, *The Principles*, 2:408–9.

28. Hermann Lotze, *Medicinische Psychologie oder Physiologie der Seele* (Leipzig, Germany: Weidman, 1852).

29. Colwyn Trevarthen, Theano Kokkinaki, Geraldo Fiamenghi Jr., "What Infants' Imitations Communicate: With Mothers, with Fathers, with Peers," in *Imitation in Infancy*, ed. Nadel and Butterworth, 127–85.

30. Paul Guillaume, *L'imitation chez l'enfant* (Paris: Alcan, 1925).

31. Henri Wallon, *De l'acte à la pensée* (Paris: Flammarion, 1942).

32. Nadel and Butterworth, "Immediate Imitation," 2.

33. Piaget and Inhelder, *The Psychology of the Child* (New York: Basic Books, 1969).

34. See Jean-Michel Oughourlian, "From Universal Mimesis to the Self Formed by Desire" and Andrew Meltzoff, "Out of the Mouths of Babes: Imitation, Gaze, and Intentions in Infant Research—the 'Like Me' Framework," both in this volume, for further commentary concerning misconceptions of imitation in modern psychology.

35. Fleming, *René Girard*, 153.

36. James G. Williams, ed., *The Girard Reader* (New York: Crossroad, 1996); Richard Golsan, *René Girard and Myth: An Introduction* (London: Routledge, 1993); Fleming, *René Girard: Violence and Mimesis*; Michael Kirwan, *Discovering Girard* (Cambridge, MA: Cowley Publications, 2005).

37. René Girard, *Deceit, Desire, and the Novel: Self and Other in Literary Structure*, trans. Yvonne Freccero (Baltimore: Johns Hopkins University Press, 1965); René Girard, *Mensonge romantique et vérité romanesque* (Paris: Grasset, 1961).

38. Girard has also published books dedicated exclusively to the works of Shakespeare: *A Theatre of Envy: William Shakespeare* (New York: Oxford University Press, 1991); and on Dostoevsky: *Resurrection from the Underground: Feodor Dostoevsky*, trans. James G. Williams (New York: Crossroads Press, 1997), a translation of *Dostoievski: Du double à l'unité* (Paris: Plon, 1963).

39. René Girard, *Things Hidden since the Foundation of the World*, trans. Stephen Bann and Michael Metteer (Stanford, CA: Stanford University Press, 1987), 8.

40. For this reason Girard has chosen to use the word mimesis instead of imitation, since imitation has been saturated by this particular cultural meaning.

41. Potolsky, *Mimesis*, 5.

42. For further commentary on the mimetic nature of desire, see Jean-Michel Oughourlian, "From Universal Mimesis to the Self Formed by Desire," Paul Dumouchel, "Emotions and Mimesis," and Jean-Pierre Dupuy, "Naturalizing Mimetic Theory," all in this volume.

43. See also Jean-Pierre Dupuy, "Naturalizing Mimetic Theory," this volume, for a discussion on the morphogenetic nature of mimesis.

44. Girard, *Things Hidden*, 8.

45. See René Girard, "Mimesis and Science: An Interview with René Girard," this volume, for additional commentary on the history of imitation.

46. Girard, *Deceit, Desire, and the Novel*; René Girard, *La violence et le sacré* (Paris: Grasset, 1972); René Girard, *Violence and the Sacred*, trans. Patrick Gregory (Baltimore: Johns Hopkins University Press, 1977); Girard, *Things Hidden*.

47. Girard, *Violence and the Sacred*; Girard, *Things Hidden*.

48. Girard, *Violence and the Sacred*; Girard, *Things Hidden*; René Girard, *Le bouc émissaire* (Paris: Grasset, 1982); and René Girard, *The Scapegoat*, trans. Yvonne Freccero (Baltimore: Johns Hopkins University Press, 1986).

49. René Girard, "Violence and Religion: Cause or Effect?" *Hedgehog Review* 6, no. 1 (2004): 8–13.

50. Girard, *Violence and the Sacred*; Girard, *The Scapegoat*.

51. See Mark R. Anspach, "Imitation and Violence: Empirical Evidence and the Mimetic Model" and René Girard, "Mimesis and Science: An Interview with René Girard," both in this volume, for a more detailed overview of Girard's anthropological theory.

52. Girard, *Violence and the Sacred*; Girard, *Things Hidden*; Girard, *The Scapegoat*; René Girard, *I See Satan Fall Like Lightning*, trans. James G. Williams (Maryknoll, NY: Orbis Books, 2001).

53. Girard, *Things Hidden*; Girard, *The Scapegoat*; Girard, *I See Satan Fall Like Lightning*.

54. See René Girard, "Mimesis and Science: An Interview with René Girard," in this volume, for additional commentary on his interpretation of the Judeo-Christian scriptures, including the influence of other religious traditions.

55. Nadel and Butterworth, eds., *Imitation in Infancy*, 1.

56. Andrew Meltzoff and M. Keith Moore, "Imitation of Facial and Manual Gestures by Human Neonates," *Science* 198 (1977): 75–78; Andrew Meltzoff and M. Keith Moore, "Newborn Infants Imitate Adult Facial Gestures," *Child Development* 54 (1983): 702–9; Andrew Meltzoff and M. Keith Moore, "Imitation in Newborn Infants: Exploring the Range of Gestures Imitated and the Underlying Mechanisms," *Developmental Psychology* 25 (1989): 954–62; Andrew Meltzoff and M. Keith Moore, "Early Imitation within a Functional Framework: The Importance of Person Identity, Movement, and Development," *Infant Behavior and Development* 15 (1992): 479–505; Andrew Meltzoff, and M. Keith Moore, "Imitation, Memory, and the Representation of Persons," *Infant Behavior and Development* 17 (1994): 83–99.

57. Andrew Meltzoff and M. Keith Moore, "Explaining Facial Imitation: A Theoretical Model," *Early Development and Parenting* 6 (1997): 179–92.

58. Andrew Meltzoff and M. Keith Moore, "Persons and Representation: Why Infant Imitation is Important for Theories of Human Development," in *Imitation in Infancy*, ed. Nadel and Butterworth, 17; Andrew Meltzoff and M. Keith Moore, "Explaining Facial Imitation," *Early Development and Parenting* 6 (1997). For a recent review of the literature on infant imitation from a variety of authors, see Nadel and Butterworth, eds., *Imitation in Infancy*; and Meltzoff and Prinz, eds., *The Imitative Mind*.

59. Colwyn Trevarthen, "Descriptive Analyses of Infant Communicative Behaviour," in *Studies in Mother-Infant Interaction*, ed. H. R. Schaffer (London: Academic Press, 1977); Colwyn Trevarthen, "Communication and Cooperation in Early Infancy: A Description of Primary Intersubjectivity," in *Before Speech: The Beginning of Interpersonal Communication*, ed. Margaret Bullowa (Cambridge: Cambridge University Press, 1979).

60. J. Nadel, C. Guerini, A. Peze, and C. Rivet, "The Evolving Nature of Imitation as a Format for Communication," in *Imitation in Infancy*, ed. Nadel and Butterworth, 209–34; J. Nadel, "Imitation and Imitation Recognition: Functional Use in Preverbal Infants and Nonverbal Children with Autism," in *The Imitative Mind*, ed. Meltzoff and Prinz, 42–62. See also Trevarthen et al., "What Infants' Imitations Communicate."

61. See Meltzoff and Moore, "Imitation, Memory, and the Representation of Persons"; Jens Asendorpf, "Self-awareness, Other-awareness, and Secondary Representation," in *The Imitative Mind*, ed. Meltzoff and Prinz; A. N. Meltzoff, "'Like Me': A Foundation for Social Cognition," *Developmental Science* 10 (2007): 126–34.

62. Michael Tomasello, A. C. Kruger, and H. H. Ratner, "Cultural Learning," *Behavioral and Brain Sciences* 16 (1993): 495–552.

63. Andrew Meltzoff, "Immediate and Deferred Imitation in Fourteen- and Twenty-four-month-old Infants," *Child Development* 56 (1985): 62–72; Andrew Meltzoff, "Infant Imitation after a 1-Week Delay: Long-term Memory for Novel Acts and Multiple Stimuli," *Developmental Psychology* 24 (1988): 470–76; Andrew Meltzoff, "Infant Imitation and Memory: Nine-month-olds in Immediate and Deferred Tests." *Child Development* 59 (1988): 217–55.

64. Jean Piaget, *Play, Dreams, and Imitation in Childhood* (New York: Norton, 1962).

65. Andrew Meltzoff and Keith Moore, "Imitation, Memory, and the Representation of Persons," *Infant Behavior and Development* 17 (1994): 83–99.

66. J. P. Klein and Andrew Meltzoff, "Long-term Memory, Forgetting, and Deferred Imitation in 12-month-old Infants," *Developmental Science* 2 (1999): 102–13; Andrew Meltzoff, "What Infant Memory Tells Us about Infantile Amnesia: Long-term Recall and Deferred Imitation," *Journal of Experimental Child Psychology* 59 (1995): 497–515.

67. Michael Tomasello et al., "Cultural Learning," 495–552; Andrew Meltzoff, P. K. Kuhl, J. Movellan, and T. J. Sejnowski, "Foundations for a New Science of Learning," *Science* 325 (2009): 284–88. See Andrew N. Meltzoff, "Out of the Mouths of Babes: Imitation, Gaze, and Intentions in Infant Research—the 'Like Me' Framework," and Ann Cale Kruger, "Imitation, Communion, and Culture," both in this volume, for a more thorough overview of the developmental research.

68. D. G. Premack and G. Woodruff, "Does the Chimpanzee Have a Theory of Mind?" *Behavioral and Brain Sciences* 1 (1978): 515–26.

69. Trevarthen et al., "What Infants' Imitations Communicate: With Mothers, with Fathers, with Peers," 140.

70. See Andrew N. Meltzoff, "Out of the Mouths of Babes," and Vittorio Gallese, "The Two Sides of Mimesis," in this volume, for their respective views on the nature of ToM based on contemporary imitation research. See also Jean-Pierre Dupuy, "Naturalizing Mimetic Theory," concerning the more implicit ToM found in Girard's mimetic theory.

71. For a recent review of the literature on imitation in nonhuman animals from a variety of authors, see Susan Hurley and Nick Chater, eds., *Perspectives on Human Imitation: From Neuroscience to Social Science*, vols. 1 and 2 (Cambridge, MA: MIT Press, 2005).

72. Michael Tomasello and Josep Call, *Primate Cognition* (New York: Oxford University Press, 1997); R. Bryne, "Seeing Actions As Hierarchically Organized Structures: Great Ape Manual Skills," in *The Imitative Mind*, ed. Meltzoff and Prinz, 122–40); A. Whiten, "The Imitator's Representation of the Imitated: Ape and Child," in Meltzoff and Prinz, 98–121.

73. Richard Byrne and A. Russon, "Learning by Imitation: A Hierarchical Approach," *Behavioral and Brain Sciences* 21 (1998): 667–721.

74. Richard Byrne, "Detecting, Understanding, and Explaining Imitation in Animals," in *Perspectives on Human Imitation*, ed. Hurley and Chater, 1:225.

75. See Ann Cale Kruger, "Imitation, Communion, and Culture," in this volume, for a more detailed review of the comparative research on primate imitation and cognition.

76. Michael Tomasello, *The Cultural Origins of Human Cognition* (Cambridge, MA: Harvard University Press, 2000); Merlin Donald, *Origins of the Modern Mind: Three Stages in the Evolution of Culture and Cognition* (Cambridge, MA: Harvard University Press, 1991).

77. Merlin Donald, *A Mind So Rare* (New York: W.W. Norton & Co., 2001), 263.

78. For a recent review of research on mirror neurons and their functional and evolutionary significance from a variety of authors, see M. Stamenov and Vittorio Gallese, *Mirror Neurons and the Evolution of Brain and Language* (Amsterdam: John Benjamins Publishing Co., 2002); and Hurley and Chater, *Perspectives on Human Imitation*.

79. G. Rizzolatti, L. Fadiga, L. Fogassi, and V. Gallese, "Premotor Cortex and the Recognition of Motor Actions," *Cognitive Brain Research* 3 (1996): 131–41; Vittorio Gallese, L. Fadiga, L. Fogassi, G. Rizzolatti, "Action Recognition in the Premotor Cortex," *Brain* 119 (1996): 593–609.

80. G. Rizzolatti, L. Fogassi, and V. Gallese, "Neurophysiological Mechanisms Underlying the Understanding and Imitation of Action," *Nature Reviews Neuroscience* 2, no. 9 (2001): 663.

81. G. Rizzolatti and M. Arbib, "Language within Our Grasp," *Trends in Neuroscience* 21, no. 5 (1998): 188–94; O. Gruber, "The Co-evolution of Language and Working Memory Capacity in the Human Brain," in Stamenov and Gallese, *Mirror Neurons and the Evolution of Language*, 77–86.

82. Gerhard Roth, "Is the Human Brain Unique?" in *Mirror Neurons and the Evolution of Language*, ed. Stamenov and Gallese.

83. See Vittorio Gallese, "The Two Sides of Mimesis," in this volume, for a more detailed review of the research on mirror neurons and its significance to understanding primate cognition and sociality.

84. For a recent review of the literature on the neuroscience of human imitation from a variety of authors, see Meltzoff and Prinz, *The Imitative Mind*; Hurley and Chater, *Perspectives on Human Imitation*.

85. M. Iacoboni, R. Woods, M. Brass, H. Bekkering, J. Mazziotta, G. Rizzolatti, "Cortical Mechanisms of Human Imitation," *Science* 286, no. 5449 (1999): 2526–28.

86. G. Rizzolatti et al., "Neurophysiological Mechanisms Underlying the Understanding and Imitation of Action," *Nature Neuroscience Reviews* 2 (2001): 661–70.

87. Marco Iacoboni, "Understanding Others: Imitation, Empathy, and Language," in *Perspectives on Human Imitation*, ed. Hurley and Chater, 1:77–99.

88. P. Ruby and J. Decety, "Effect of Subjective Perspective Taking during Simulation of Action: A PET Investigation of Agency," *Nature Neuroscience* 4 (2001): 546–50; J. Decety, "Is There Such a Thing As a Functional Equivalence between Imagined, Observed, and Executed Action?" in *The Imitative Mind*, ed. Meltzoff and Prinz, 291–310; J. Decety and T. Chaminade, "Neural Correlates of Feeling Sympathy," *Neuropsychologia* 41 (2003): 127–38. See also P. L. Jackson, A. N. Meltzoff, and J. Decety, "How Do We Perceive the Pain of Others: A Window into the Neural Processes Involved in Empathy," *NeuroImage* 24 (2005): 771–79; C. Lamm, A. N. Meltzoff, and J. Decety, "How Do We Empathize with Someone Who Is Not Like Us? A Functional Magnetic Resonance Imaging Study," *Journal of Cognitive Neuroscience* 22 (2010): 362–76.

89. J. Decety and J. Sommerville, "Shared Representations between Self and Other: A Social Cognitive Neuroscience View," *Trends in Cognitive Sciences* 7, no. 12 (2003): 527–33; A. N. Meltzoff and J. Decety, "What Imitation Tells Us about Social Cognition: A Rapprochement between Developmental Psychology and Cognitive Neuroscience," *Philosophical Transactions of the Royal Society: Biological Sciences* 358 (2003): 491–500.

90. Decety and Sommerville, "Shared Representations between Self and Other," 532.

91. Bruno Wicker, Christian Keysers, Jane Plailly, Jean-Pierre Royet, Vittorio Gallese, and Giacomo Rizzolatti, "Both of Us Disgusted in My Insula: The Common Neural Basis of Seeing and Feeling Disgust," *Neuron* 40 (2003): 655–64.

92. C. Keysers, B. Wicker, V. Gazzola, J. L. Anton, L. Fogassi, and V. Gallese, "A Touching Sight: SII/PV Activation during the Observation and Experience of Touch," *Neuron* 42 (2004): 335–46.

93. M. Iacoboni et al., "Grasping the Intentions of Others with One's Own Mirror Neuron System," *PLOS Biology* 3 (2003): 529–35.

94. For a recent review of the literature on the nature and mechanisms of unconscious processes, including imitation, see Ran Hassin, James Uleman, and John Bargh, *The New Unconscious* (New York: Oxford University Press, 2005).

95. A. Dijksterhuis and J. Bargh, "The Perception-Behavior Expressway: Automatic Effects of Social Perception and Social Behavior," in M. Zanna, ed., *Advances in Experimental Social Psychology* 33 (2001) (New York: Academic Press, 2001), ?

96. See T. Chartrand, W. Maddux, and J. Lakin, "Beyond the Perception-Behavior Link: The Ubiquitous Utility and Motivational Moderators of Nonconscious Mimicry," in Hassin, Uleman, and Bargh, *The New Unconscious*, 334–61.

97. J. Bargh and T. Chartrand, "The Unbearable Automaticity of Being," *American Psychologist* (July 1999): 462–79.

98. Dijksterhuis and Bargh, "The Perception-Behavior Expressway"; A. Dijksterhuis, "Why We Are Social Animals: The High Road to Imitation as Social Glue," in *Perspectives on Human Imitation*, ed. Hurley and Chater, 2: 208.

99. Dijksterhuis, "Why We Are Social Animals."

100. Ibid.

101. J. Bargh, M. Chen, and L. Burrows, "The Automaticity of Social Behavior: Direct Effects of Trait Concept and Stereotype Activation on Action," *Journal of Personality and Social Psychology* 71 (1996): 230–44.

102. A. Dijksterhuis and A. van Knippernberg, "The Relation between Perception and Behavior, or How to Win at a Game of Trivial Pursuit," *Journal of Personality and Social Psychology* 74 (1998): 865–77.

103. J. Bargh, P. Gollwitzer, A. Lee-Chai, K. Barndollar, and R. Trotschel, "The Automated Will: Nonconscious Activation and Pursuit of Behavioral Goals," *Journal of Personality and Social Psychology* 81 (2001): 1014–27.

104. A. Dijksterhuis, "Why We Are Social Animals."

105. Marcel Kinsbourne, "Imitation as Entrainment: Brain Mechanisms and Social Consequences," in *Perspectives on Human Imitation*, ed. Hurley and Chater, 2:163–72.

106. J. Bargh, "The Most Powerful Manipulative Messages Are Hidden in Plain Sight," *Chronicle of Higher Education*, January 29, 1999, B6.

107. J. M. Oughourlian, *The Puppet of Desire: The Psychology of Hysteria, Possession, and Hypnosis* (Stanford, CA: Stanford University Press, 1982); Jean-Michel Oughourlian, *The Genesis of Desire*, trans. Eugene Webb (East Lansing: Michigan State University Press, 2010); Webb, *The Self Between*; Scott Garrels, "Imitation, Mirror Neurons, and Mimetic Desire: Convergent Support for the Work of René Girard," *Contagion: Journal of Violence, Mimesis, and Culture* 12–13 (2006): 47–86.

108. Oughourlian, in *The Puppet of Desire* and *The Genesis of Desire*. See Webb, *The Self Between*; William Hurlbut, "Mimesis and Empathy in Human Biology," *Contagion: Journal of Violence, Mimesis, and Culture* 4 (1997): 14–25; Garrels, "Imitation, Mirror Neurons, and Mimetic Desire."

109. Webb, *The Self Between*, 213.

110. Paisley Livingston, *Models of Desire: Rene Girard and the Psychology of Mimesis* (Baltimore: Johns Hopkins University Press, 1992), xvi.

111. Hurley and Chater, *Perspectives on Human Imitation*, 1:48.

112. Meltzoff and Prinz, *The Imitative Mind*.

113. Hurley and Chater, *Perspectives on Human Imitation*, 50–51.

114. Girard, "Mimesis and Violence," *Berkshire Review* 14 (1979): 9, reprinted in *The Girard Reader*, ed. James G. Williams (New York: Crossroads Publishing Company, 1996), 9–19.

115. For a recent review, see John Eldridge, "What Effects Do the Treatment of Violence on the Mass Media Have on People's Conduct? A Controversy Re-considered," in *Perspectives on Imitation*, ed. Hurley and Chater, 2:243–55; and L. Rowell Huesmann, "Imitation and the Effects of Observing Media Violence on Behavior," in *Perspectives on Human Imitation*, ed. Hurley and Chater, 2:257–66.

116. Williams, *The Girard Reader*, 4.

117. Girard, "Violence and Religion: Cause or Effect?" 18–19.

Part 1

Imitation in Child Development
and Adult Psychology

CHAPTER 2

From Universal Mimesis to the Self Formed by Desire

Jean-Michel Oughourlian

Universal Mimesis

Martin Heidegger began his course in the summer semester of 1935 with the fundamental question of metaphysics: Why are there beings rather than nothing? I can do no better in introducing this chapter than to imitate him by posing what seems to me the essential question today in psychology: Why is there *movement* rather than nothing? How does one teach a small child to say "Papa" or "Mama" or "cookie"? How does one teach him to speak the language of those around him? Answering these questions does not require complicated experiments. There is no need for measuring instruments or for statistical calculation. Immediate observation, everyday experience, and plain good sense are sufficient to answer them. All one need do is repeat the word to the child enough times so that he repeats it. The procedure works perfectly, as everyone knows, and there is no other, known or imaginable, that could produce the same result.

Why is the child moved to repeat a word we pronounce in his presence? The problem posed here is that of identifying what enables learning to take place. The essential condition is obviously that the child's interest must be attracted, and his attention held, by the adult. Only that interest, that attention, that attraction are capable of arousing in the child the effort needed to reproduce the proposed phoneme, or indeed to learn anything at all. Is that sequence of events normal? Yes! Is it uniform? No! To see that it is not, one need only visit a center for child psychiatry to observe its difficulty in autistic children.

What is the force that, from the beginning of life, draws the child into reproducing what an adult says or does? This force of attraction, interest, and attention is so much a part of the fabric of humanity that it is taken for granted. A young child has no power to resist that attraction. To feel such attraction is the child's very nature, to the degree that he or she is "normal." A child lacking this capacity would be deprived of something basic to his humanity; he would become isolated, autistic. That natural force of cohesion, which alone grants access to the social, to language, to culture, and indeed to humanness itself, is simultaneously mysterious and obvious, hidden in and of itself, but dazzling in its effects—like gravity and the attraction of corporeal masses in Newtonian space. If gravity did not exist, life on earth would be impossible. Similarly, if this remarkable force that attracts human beings to one another, that unites them, that enables children to model themselves on adults, that makes possible their full ontogenesis and, as I just said, their acquisition of language—if this force did not exist, there would be no humanity.

This force, as fundamental for psychology as gravity is for physics, is what, following René Girard, I call *mimesis*. The attraction the child feels toward the adult, the attention he gives to what one says or does, this essential factor in all education is one with the force I am attempting to specify. The mechanism of learning—that is, imitation or repetition—is the objective translation of mimesis, its expression in activity.

Taking as a model the theory of universal gravitation in physics, let us propose the hypothesis that there is a single principle at the foundation of all the human sciences: *universal mimesis*. In both psychology and sociology, the most basic and elementary manifestation of this principle is the force of attraction that draws people together and determines their interest in one another. Should we wish to pursue this analogy to physics, it could be said that the mimesis between two individuals is the force of attraction that each simultaneously exerts on the other and submits to. This force is proportional to the mass, as it were, of each, and inversely proportional to the distance between them.

What, however, is it that one can refer to as "mass" in psychology? For a young child in his or her relation to an adult, the notion of "mass" can be interpreted almost literally: the mass of the adult in comparison with the child's helps explain the child's tendency to seek and submit to the adult's influence. Between adults, there is also a force of attraction, but in this case the notion of "mass"—which is to say, that which each represents for the other and the capacity each has to influence or attract the other—becomes more complex. Mass correlates closely with quantity. The mimesis that a crowd triggers, the power of influence a group has, is proportional to the number of individuals in it. It is this prodigious magnification of the force of mimesis that explains the difference between the psychology of individuals and group psychology, and the stupendous transformations that the former can undergo when influenced by the mimetic power of a crowd or mob.

Humans are social animals; it is our relation to the other that makes us human, and this relation is mimetic in the sense that only the play of that force makes our humanity possible. But mimesis is not only a force of attraction; it is also a force of repulsion. Imitation begins as discipleship, in which the model is taken simply as a model—but before long, the imitation of a gesture will cause the model and the disciple to grasp at the same object: the model will become a rival, and mimesis will take on the character of conflict. In this way, mimesis engenders both attraction and repulsion; in this way it produces discipleship and conflict, nonviolent and violent acquisition, peace and war, alliance and tension, likeness and difference.

Jean Piaget was the principal psychologist of intelligence, Sigmund Freud the psychologist of affectivity. In reality, however, the separation of these two areas of psychic life is arbitrary and academic: both have their source in mimesis. Piaget and Freud sought to deconstruct the mechanisms of one or the other function by returning to a point of lesser complexity, but without ever reaching their root and discovering the underlying simplicity of their shared origin. Piaget's genetic epistemology is the history of the subject's cognitive adaptation to his milieu. It distinguishes six stages in that process. Freudian psychoanalysis traces a different set of stages in the evolution and progressive adaptation of affectivity to its milieu. Neither method inquires about the essential first condition for the rise of the mechanisms and operations it distinguishes.

Contemporary experimental psychology, as it has refined its methods and experiments, has made significant strides in approaching my point of view. The work of Meltzoff and Moore is seminal and has shown that newborns imitate manual gestures and facial expressions.[1] In carefully controlled experiments, these newborns in fact stuck out their tongues and opened their mouths in imitation of the experimenter. Piaget always claimed that this type of imitation does not appear until around the age of 8 to 12 months. In the same experiments, Meltzoff and Moore obtained from their newborns a *deferred* imitation; that is, as they describe it in their paper, the baby's reproduction of adult gestures (such as tongue protrusions) took place *after* the gesture enacted by the experimenter had vanished from its field of view for delays up to 24 hours. For Piaget, deferred imitation came first and was a manifestation of the symbolic function, that is, of the possibility of representing to oneself information about objects that are absent. I am suggesting here, in agreement with Meltzoff and Moore, the opposite view: it is imitation, immediate and then increasingly deferred, that little by little, by its very process, constitutes representation, the symbolic function, and consciousness itself, with all its attributes.

I want to underline the precociousness of the attraction exercised by the adult, and the influence undergone by the child. At this very early stage, the mimetic force works alone and is the sole cause of the reproduction of the gesture. With regard to the *cognitive* level, Piaget is correct in thinking that at this stage it is much too early for the child to form conscious mental representations. Nevertheless, experiments prove that imitation takes place. On the *affective* level, the gesture of sticking out the tongue, for example, has no significance at all for the baby: mimesis therefore precedes and gives rise to meaning and the symbolic function; it is not the result, but the condition. It is only much later that the child will discover the affective and relational value of reproducing the sign or gesture. Only later still will he or she discover the affective or cultural connotations of the gesture. At the root of all the discoveries the child will make, at the starting point of its entry into culture, one finds mimesis, the force by which the child is drawn into the system of relations. It is only on the basis of that fundamental given that one can study the mechanisms and stages of cognitive development that mimesis produces.

But how does that mimetic force operate or get brought into operation? What sorts of neurological or neurophysiological systems are indispensable to its operation? These questions pertain to neurophysiology, and perhaps also to biochemistry. The discovery of "mirror neurons"—first identified in monkeys,[2] but also thought to exist in the human brain based on neuroimaging studies[3]—may very well be one such neurophysiological substrate for the operations of mimesis. These brain cells are activated both when the subject watches an action being performed and when the subject performs the action. If I observe someone reaching for an object, the same motor neurons are activated "as if" I am performing the action myself. Observing the gesture puts me in the other person's place and even prepares me to execute the same action.

Mirror neurons are activated not by just any behavior or movement, but rather by intentional or goal-directed behavior. Additional experiments from developmental psychology further illustrate how imitation goes beyond surface behaviors to underlying goals and intentions.[4] Children watched adult experimenters trying and failing to pull apart an object. Rather than imitating the failed attempt modeled by the experimenter, the children instead tried various strategies for accomplishing the underlying goal, even though they never saw the experimenter successfully accomplish this himself.

The "psychological movement" that the adult's suggestion brings about in the child has an *intentioned teleology*. This phenomenon—the imitation of goals and aims, intentions and desires—is what Girard has called *mimetic desire*. The above research on infant imitation and mirror neurons resonates suggestively with the intuitions that Girard expressed 50 years ago in his first book, *Deceit, Desire, and the Novel*. There he explains that the greatest writers succeed in understanding and describing the mechanisms of imitation that are continuously taking place at a level below conscious awareness, drawing us into the orbit of one another's desires in increasingly complex ways.

Girard, however, takes an additional step that scientists interested in imitation have not yet taken: he links goal-oriented imitation to conflict. Indeed, he posits that imitation is at the root of all the unique forms of human rivalry and violence. Understanding that imitation not only underlies positive human abilities, such as empathy or learning, but also presents terrible dangers is critical to understanding Girard's anthropological theory with respect to the origins of human culture and religion in scapegoating violence.[5] In truth, it does not take a very big deductive leap to extrapolate the potentially dangerous consequences of the imitation of goals.

In an interview with the French daily *Le Figaro*, neuroscientist Giacomo Rizzolatti, leader of the team that discovered mirror neurons, remarked that "the process of imitation is limited in monkeys, and it's often dangerous for them to imitate."[6] What is true for monkeys holds as well for humans. The person who takes me as a model confirms and reinforces my desire. Such reinforcement can help friendship grow and flourish. Rivalry occurs, however, when desires converge mimetically on an object that cannot be shared. If two friends fall in love with the same girl, the very same reciprocity that brought them together could tear their friendship apart. Nor is such rivalry limited to erotic competition. It emerges whenever mutually thwarted desires reinforce each other in an escalating spiral: colleagues competing for a job or a fellowship, strangers for a parking spot, politicians for the leadership of a political party, neighboring countries for a geographic territory, and so on. The more each perceives the other to desire the object, the more desirable the object becomes.

THE SELF FORMED BY DESIRE

The imitation of the desire to possess an object, or *appropriative mimesis*, as Girard and I termed it, does not suffice to produce a self.[7] The building up of a self requires a type of imitation that bears on the very *being* of the model. This type of imitation, which serves to ontologize the self, coincides to a large degree with what Freud termed "identification." Once this psychological process is set in motion, it is unceasing—because, as I wish to emphasize, humanity is entirely in the hands of mimesis, the free play of which is not curtailed by any "instinct," not trammeled by anything issuing from the genetic dimension of mimesis.

Appropriative mimesis brings with it, therefore, a tendency to rivalry that cannot be resolved by any sort of "dominance pattern" or instinctual schema for the ritualization of conflicts such as those operating among animals. A noninstinctual solution for this rivalry is needed—a psychosociological solution. With regard to the societal aspect, Girard has argued persuasively that what serves to put a stop to the general violence proceeding from appropriative mimesis is the mechanism of victimization.[8] With regard to the psychological and individual aspect, the mimetic conflict unleashed by appropriative mimesis has a history that blends with that of psychogenesis. When a solution cannot be found in the physical death of the model—owing to the weakness of the child and the power of the societal structures (that is, culture) deriving from prior victimizations—there remains only one possible form of victory in the struggle for psychic life: the psychological "death" of the model. This means the appropriation by the child of the very *being* of the model through the process of identification. Freud had the genius to notice this mechanism, which amounts to the foundation of psychogenesis and the formation of personality, but encumbered as he was by mythology and the idea of instincts, he failed to realize how purely it was a function of mimesis. It is clear that the models that perish most often in this way are those the child has the most constant involvement with: his parents. But I maintain that identification is an abiding mechanism for the resolution of mimetic conflict; that consequently, the number of times it operates in a life is practically unlimited; and finally, that it is a *physiological* mechanism.[9] Every other device by which we seek to lay claim to what another has, that is, to affirm the precedence of our own desire over that of the other, ends in a *pathological* process.

I have always thought that what one customarily calls the "I" or "self" in psychology is an unstable, constantly changing, and ultimately evanescent structure. I think, to evoke the intuitions of Hegel on this point, that only *desire* brings this self into existence. Because desire is the only psychological motion, it alone, it seems to me, is capable of producing the self and breathing life into it. The first hypothesis that I would like to formulate in this regard is this: *desire gives rise to the self and, by its movement, animates it.* The second hypothesis, which I have adopted unreservedly since I first became aware of it, is that *desire is mimetic.* This postulate, which was advanced by Girard as early as 1961, seems to me capable of serving as the foundation for a new, pure psychology—that is, one unencumbered by any sort of biologism.[10]

Desire is the source of the self. The self is thus, in fact, a self of desire. Because the self is engendered by desire, it cannot lay claim to the ownership of that desire. Furthermore, desire is mimetic, since it reproduces or copies another desire. Therefore the self cannot claim that the desire that constitutes it has priority over another's desire.

These two hypotheses make it necessary to revise earlier psychologies, since these are psychologies either of the subject or of the object. They demand that one renounce the mythical claim to a self that would be a permanent structure in a monadic subject.[11] A new psychology developed on the basis of these premises will have to be a psychology of movement and change. It will not be a psychology of the individual; rather it will be *interdividual.*

Those psychologies referred to as "psychologies of the subject" are of great diversity. They are pertinent to the questions that interest us in that they suppose that the subject is a delimited and independent psychological entity, and that psychological movement is born in the inner depths of that entity—that the source of dynamism lies within the subject. Freudian psychoanalysis, which is the best-known form of subjective psychology, calls this dynamic force "libido." From the time of his birth, and throughout the course of his life, the Freudian subject must "invest" this energy—capital surging up from his own biological mass. Prime targets for such investments are other subjects, but these carry so little weight in the process by which libidinal investment is determined or in its magnitude that Freud calls them "objects." As Tran-Thong puts it: "In Freud's scheme of object relations, persons do not seem to have a positive influence on the child's affectivity; they are only inert objects of the libido, and they can be of any sort. Moreover, the development of one's relations to objects is unilaterally determined by the development of the libido."[12] And Piaget says: "The Freudian instinct is a sort of substantial force or energy that conserves itself as such while transferring itself from one object to another."[13] According to the Freudian conception, the energy that wells up in this way has to flow off somewhere, and the flowing is accompanied by pleasure. But it also often has to be held back, with the result that it collides with reality even as it emerges. Energy that becomes compressed in this way will produce tensions and seek outlets, either in the psychological direction (neurotic symptoms) or in the organic direction (hysterical displacements and psychosomatic symptoms).

The Freudian theory of the instinctual origin of psychological movement seems to me contrary to reality and to observed psychological facts. Those facts exist precisely because humans have become freed, in the course of a long phylogenesis, from the power of instinct. In no case can a psychological fact be the expression of an instinct.[14] To the contrary: the psychological bears witness, each time it appears, to the freedom that characterizes human behavior in comparison with the instinctive. By the logic of the mimetic perspective, and in deference to the reality of facts, subject psychologies and subjectivism must be rejected: to suppose that psychological movement originates within the interior of a "monadic" subject is a mythical illusion.

More serious and dangerous still is the approach of those psychologists who place the origin of psychological movement totally outside the subject, in the object (interpreted in an instrumental sense and not at all in the sense that psychoanalysis attributes to it). In fact, the object as such possesses no force at all, either of attraction or of determinism, and it cannot be turned into a stimulus eliciting a psychological response—except under experimental conditions where the complex interactions that are introduced are the real cause of the effects produced but are ignored by behaviorists.

As Arthur Koestler emphasized, there is nothing in the pompous titles of the works of B. F. Skinner—*The Behavior of Organisms* and *Science and Human Behavior*—to indicate that the data used come entirely from experiments on pigeons and rats.[15] The attempts of behaviorists to pass from the Skinner box and the rat to human beings lead them to radical theories that have no relation to the simplest observations of actual people. Are we really to believe that human reactions are fortuitous, totally random, and that if a subject reacts in a certain way, it is simply the consequence of "reinforcement," the effect of his having been rewarded in some way by the result obtained? This is a claim that is all the more appealing to behaviorists in that the reinforcements themselves are founded, according to their thinking, on the satisfaction of instincts or needs, something that once again roots the source of psychological movement in the biological. These object psychologies blithely carry over forms of animal behavior into the human, completely obscuring what is fundamental: the progressive liberation of man from instinctual imperatives in the course of his phylogenesis, the development of what we call human freedom, even if it often degenerates into servitude and unhappiness.[16]

Interdividual Psychology

Humanity, as Aristotle said, is distinguished from all other animals by our mimetic capacity: "Imitation is natural to man from childhood, one of his advantages over the lower animals being this, that he is the most imitative creature in the world, and learns at first by imitation."[17] There is, between humans and other animals, a radical epistemological gulf that seems to have been overlooked completely by the behaviorists, and partially by the psychoanalysts, because of their insistence on rooting psychological movement in the biological. Psychological movement is, from the beginning of human life, mimetic. It is mimesis, and that alone, that makes one human, that constitutes the self, and that makes possible one's entry into the sphere of language. This means that from the very start, psychological actuality is to be found *between* individuals. This is why Rene Girard, Guy Lefort, and I were led to propose a pure psychology, for which we could find no better name than *interdividual psychology*.[18]

Henri Wallon (1879–1962) was the first psychologist to acknowledge all that the self owes to the other, and to maintain that the fundamental mechanism of psychic development is one of a complementary duality—a duality that he formulated, in the domain of the person, as a dialectic of the self and the other. As Tran-Thong says in his comments on Wallon: "In the beginning the other is simply the social nature of man. The newborn, in his utter helplessness, is a biologically incomplete and insufficient being who requires society in order to gain existence, who calls on the other to complete him. The self and the other are thus bound together in a fundamental way at the point of origin by a tie that is ontological and existential. . . . The genesis of the self cannot take place except by the mediation of the other and simultaneously with the other in a process of differentiation that is gradual and reciprocal."[19] In reading this text, one cannot help thinking that Wallon was very close to our interdividual psychology. Still, he lacked the idea of mimesis as the moving force in the dialectical couple "self-other." Psychological movement is touched off at the time of the linking of the two psychological entities by way of the mimetic connection that enables the movement of one to be communicated to the other.

We must give due recognition to the interchangeability, the porosity, and the constant interaction between the self and the other. The self is a purely psychological entity, a structure in constant becoming at the heart of continuous exchanges with similar structures.

Figure 1. Interdividual relation between two psychological entities

Interdividual psychology is the study of the types of interaction that take place between psychological entities—interactions that actively constitute the existence and the individuality of each person. Mimesis is the principle that governs the genesis of selves and their interactions. It is a principle of the transmission of information from one self or several to one or several others.

Let us represent the relation between two psychological entities, A and B, as in the figure above. We can see that the perspective of interdividual psychology is a comprehensive one. It embraces the arrow AB, when A is taken to be the self and B the other, as well as the arrow BA, when B is the self and A the other. The true psychological actuality is the relationship *between* the two. This psychological relationship, which is exclusively mimetic, I call the *interdividual relation*.

It is time to formulate a third hypothesis, which follows from the other two: *Psychological actuality is located not in the tranquil opacity of any "body" in the strict sense, or in the reassuring wholeness of any self, but rather in the mysterious transparency of the interdividual relation.*

According to the logic of these hypotheses, the self, which is engendered by the desire that constitutes it and is in turn simply the desire of the other mimetically transposed, will in most cases maintain itself in existence by way of two forms of misrecognition, or *méconnaissance*. The self exists by misrecognizing what it owes to the desire of the other that produces and animates it. This misrecognition leads the self openly to claim the desire as its own. Desire cannot exist autonomously and develop an "I" in A except by misrecognizing how much it owes to desire B, to the desire of the other that it copies. Misrecognition at this level is functional and normal. It serves to make possible the existence of the self of desire. This misunderstanding will remain "normal" and functional to the extent that the other is taken as a *model*, that is, to the extent that the interdividual relation remains peaceful.

But the model does not always remain a model to the necessary degree. When he or she becomes a *rival*, the misrecognition becomes pathogenic because it embraces a double claim: a claim on the part of the self to *ownership* of its own desire, and a claim on the part of the desire to its *anteriority* to the desire of the other, to *priority* over the other's desire.

Misrecognition always has to do with the interdividual relation and, to be even more precise, with the mimetic character of that relation. It bears on the fact that it is B's desire that arouses or engenders A's desire; that desire A is born mimetically from its modeling on desire B, from which it derives its own existence; and that desire A in turn produces, by its own movement, self A, the self of desire with all its attributes: consciousness, memory, attention, and so on. If one were trying to develop a psychology of consciousness, this misrecognition would have to be cast as "unconscious." The self, founded on misrecognizing, could not be conscious of it. Nor could the desire that engenders that self. But this misrecognizing is a failure to understand factual reality, which is *constant* and *objective*. It should not under any circumstances be confused with the psychoanalytic unconscious, which is fundamentally *subjective*, in that it is a hidden part of the self, a part that is constituted, for each self, by its own history.

Since this interdividual mimetic relation is fundamentally misapprehended, how can one bring it into view, notice, and recognize it? To *uncover* it, in the most literal sense of the term, as I have argued most recently in *The Genesis of Desire*, is the goal of mimetic psychotherapy.[20] It seems to me that to accomplish this, one has to try to catch hold of the interdividual relation at certain special points at which it shows itself. One has to study psychological phenomena that are unusual, that are exceptional without exactly being rare, and that are capable of throwing light on one another. The presence or absence of an element in one or another of the phenomena would offer clues; perhaps such tentative linkages can put us on the path to the real facts.

To this end, in my first book, *The Puppet of Desire*, I decided to put together a phenomenology of mimetic desire.[21] I chose for this purpose some phenomena that can be clearly shown to be mimetic: magic, the casting of spells, sorcery, exorcism, adorcism, hysteria, and finally hypnosis.[22] In the process, I sketched out a sort of history of the interdividual relation, of the forms that the relation between the self and the other has taken over the centuries. This "other" may be a definite and particular other (the "little other" of Lacan or the "internal mediator" of Girard), or a cultural entity, an "Other" made up of all that the self has learned, seen, or read—of everything in which it is immersed, physically and intellectually (the "Grand Autre" of Lacan or the "external mediator" of Girard).[23] This history shows successively the "other" in all its guises: a real other, absent and maleficent in enchantment; a virtual Other, cultural and external—the Devil in certain forms of sorcery and demonic possession; a cultural Other, virtual and external, but beneficent, in African possession (adorcism); an incarnate otherness in hysteria, and integrated otherness in psychoanalysis (the hysteric disguising the interdividual relation as a physical symptom, and Freud disguising it as a psychic symptom). And finally, in a way that can demonstrate the interdividual relation experimentally, a real other, present and usually beneficent, in the phenomena of hypnosis, which undiscerning observers have beheld for centuries without understanding them.

In all my work I have tried to show that every glimmer of recognition (_reconnaissance_) of the interdividual relation is therapeutic for the person and confers on him identity and unity because it consists in taking the other consciously as a model. I have also tried to make clear that all obstinacy in the misrecognizing (_méconnaissance_) of that relation and of the mimetic character of desire impedes the development of identity, and that taking the other as a rival produces in the person a dissociation and pathology commensurate with that rivalry.

The application of the hypotheses formulated above to the interpretation of verified psychological, psychosociological, or psychopathological phenomena should prove capable not only of stimulating further study but also of moving into new possibilities of synthesis and the discovery of new connections between those phenomena and recent research on human imitation. The study of the dialectic of desire at work in each of these phenomena exposes the mimetic mechanisms that produce them, and allows us to see how each is a form of the _concealing_ and also, in a certain manner, the _revealing_ of the interdividual relation.

ACKNOWLEDGMENTS

This chapter was originally published in French in 1982 as "De la mimésis universelle au moi-du-désir," in J.-M.Oughourlian, _Un Mime nommé desir: Hystérie, transe, possession, adorcisme_ (Grasset et Fasquelle). The English translation was published in 1991 (translation by Eugene Webb) as "From Universal Mimesis to the Self Formed by Desire," in J.-M. Oughourlian, _The Puppet of Desire: The Psychology of Hysteria, Possession, and Hypnosis_ (Stanford University Press). This chapter has been revised and abridged by the author for this volume with permission from the current publishers (L'Harmattan and Stanford University Press). The translator's original comments are included in the notes in brackets.

NOTES

1. Andrew Meltzoff and Keith Moore, "Imitation of Facial and Manual Gestures by Human Neonates," _Science_ 198 (1977): 75–78; Andrew Meltzoff and Keith Moore, "Newborn Infants Imitate Adult Facial Gestures," _Child Development_ 54 (1983): 702–9; Andrew Meltzoff and

Keith Moore, "Imitation in Newborn Infants: Exploring the Range of Gestures Imitated and the Underlying Mechanisms," *Developmental Psychology* 25 (1989): 954–62. See also Andrew N. Meltzoff, "Out of the Mouth of Babes: Imitation, Gaze, and Intentions in Infant Research—The 'Like Me' Framework," in this volume.

2. Vittorio Gallese et al., "Action Recognition in the Premotor Cortex," *Brain* 119 (1996): 593–609; Vittorio Gallese, "The Manifold Nature of Interpersonal Relations: The Quest for a Common Mechanism," *Philosophical Transactions of the Royal Society of London, Series B, Biological Sciences* 358 (2003): 517–28; Giacomo Rizzolatti et al., "Premotor Cortex and the Recognition of Motor Actions," *Cognitive Brain Research* 3 (1996): 131–41; Vittorio Gallese, "The Two Sides of Mimesis: Mimetic Theory, Embodied Simulation, and Social Identification," this volume.

3. Marco Iacoboni et al., "Cortical Mechanisms of Human Imitation," *Science* 286, no. 5449 (1999): 2526–28.

4. Andrew Meltzoff, "Understanding the Intentions of Others: Re-enactment of Intended Acts by 18-Month-Old Children," *Developmental Psychology* 31 (1995): 838–50.

5. See also Mark R. Anspach, "Imitation and Violence: Empirical Evidence and the Mimetic Model," and René Girard, "Mimesis and Science: An Interview with René Girard," both in this volume.

6. *Le Figaro*, interview of 5 February 2005.

7. René Girard, *Things Hidden since the Foundation of the World*, trans. Stephen Bann and Michael Metteer (Stanford, CA: Stanford University Press, 1987).

8. See Girard, *Things Hidden*.

9. "Physiological" here means natural and healthy and refers to a process that is normal as opposed to pathological.

10. René Girard, *Deceit, Desire, and the Novel: Self and Other in Literary Structure*, trans. Yvonne Freccero (Baltimore: Johns Hopkins University Press, 1965).

11. [The concept of the mythical here has special meaning in the context of the previous writings of Oughourlian and Girard, where it refers not just to fictitiousness, but also to an imaginative pattern of antagonism in which one's own self or group is always cast in the role of the unblemished hero or protagonist while one's opponents are cast as monstrous villains. In his *Violence and the Sacred*, Girard contrasts the "mythic" in this sense with the "tragic," in which the distinctions of value between the opposing sides are less clear-cut.—Trans.]

12. Tran-Thong, *Stades et concept de stade de développement de l'enfant dans la psychologie contemporaine* (Paris: Vrin, 1967), 321.

13. Jean Piaget, *Introduction a l'épistémologie génétique*, vol. 3, *La pensée biologique, la pensée psychologique et la pensée sociologique* (Paris: Presses Universitaires de France, 1950).

14. [In accord with standard French usage, Oughourlian's "psychological" refers to human psychology; his statement here should not be interpreted as denying the legitimacy of what in English is referred to as "animal psychology," which includes the study of instinctive behavior.—Trans.]

15. Arthur Koestler, *The Ghost in the Machine* (New York: Macmillan, 1968).

16. See William B. Hurlbut, "Desire, Mimesis, and the Phylogeny of Freedom," this volume.

17. See Richard McKeon, ed., *The Basic Works of Aristotle* (New York: Random House, 1941), 1457.

18. See Girard, *Things Hidden*.

19. See Tran-Thong, *Stades et concept de stade*, 184–85.

20. Jean-Michel Oughourlian, *The Genesis of Desire*, trans. Eugene Webb (East Lansing: Mich gan State University Press, 2010).

21. Jean-Michel Oughourlian, *The Puppet of Desire: The Psychology of Hysteria, Possession, and Hypnosis*, trans. with intro. by Eugene Webb (Stanford, CA: Stanford University Press, 1991).

22. ["Adorcism" is the invoking of a possession, that is, the opposite of "exorcism," or the casting out of a possession.—Trans.]

23. [The author's comparison to Lacan and Girard could be misleading, since in some aspects the analogies to their concepts are somewhat oblique. Lacan actually uses the term "Other" in a number of ways, some of which would correspond more closely with Oughourlian's use than others (see Anika Lemaire, *Jacques Lacan* [London: Routledge and Kegan Paul, 1977],157; Ellie Ragland-Sullivan, *Jacques Lacan and the Philosophy of Psychoanalysis* [Urbana: University of Illinois Press, 1986],15–16). Lacan's term "other," spelled with a lower case "o" (*objet a*), is an object that arouses an insatiable desire, something unattainable, especially when the desire of some important person (such as one's mother) for the object indicates its desirability (see Jacques Lacan, *The Four Fundamental Concepts of Psychoanalysis* [New York: Norton, 1978], 67–122). Lacan also uses the term "other" (*autre*) for the indirect object of "empty" (distorting, concealing) speech and "Other" (*Autre*) for the source of "full" (truthful, revealing) speech, as when he says that "the unconscious is the discourse of the Other" (Jacques Lacan, *Ecrits* [Paris: Seuil,1966], 549; cf. Bice Benvenuto and Roger Kennedy, *The Works of Jacques Lacan: An Introduction* [New York: St. Martin's Press, 1986], 86–87). Girard's terms "internal mediator" and "external mediator" are from his *Deceit, Desire, and the Novel* (1965:9). The terms refer to the other conceived of as a competitor within one's own concrete world (an internal mediator) or (an external mediator) as an ideal model with whom one cannot come into competition either because he is a figure in an ideal realm (in which case the external mediator is a cultural Other in Oughourlian's sense) or because his sphere of actions does not overlap with one's own. In the latter case, the external mediator would not be a cultural entity in the same sense as Christ is for a Christian or Amadis of Gaul was for Don Quixote (two examples given by Girard), but his position as "external" to one's own sphere of possibility would be culturally defined as such. –Trans.]

CHAPTER 3

Out of the Mouths of Babes: Imitation, Gaze, and Intentions in Infant Research—the "Like Me" Framework

Andrew N. Meltzoff

In the past decade, we witnessed an overturning of one of the most pervasive scientific myths about man's original nature—the myth of the asocial infant. The classical scientific views offered by Freud, Skinner, and Piaget proposed that the newborn is at first cut off from others and gradually becomes "socialized." Freud and his followers proposed a distinction between a *physical* and *psychological* birth. When a child is born, there is a physical birth but not yet an interpersonal one. The baby is like an unhatched chick, incapable of interacting as a social being because a "barrier" leaves the newborn cut off from external reality. Freud sought a metaphor to describe the psychology of the newborn-parent relationship and likened the child's situation to the isolation found inside a shell: "A neat example of a psychical system shut off from the stimuli of the external world. . . . is afforded by a bird's egg with its food supply enclosed in its shell; for it, the care provided by its mother is limited to the provision of warmth."[1] The newborn babe was a social isolate.

Piaget's child is similar, but he uses a philosophical rather than biological metaphor. Piaget believed that the newborn is "radically egocentric" or even "solipsistic."[2] The newborn has only a few reflexes at her disposal (sucking, grasping) and develops through a series of stages, chiefly by manipulating the physical world. The child breaks free of the initial solipsism by 18 months, as described in Piaget's influential book *The Construction of Reality in the Child.*[3] It is a long and hard journey to progress from a state of solipsism to one of intersubjectivity and emotional rapport with others.

Skinner claimed that the newborn was a social tabula rasa.[4] One cannot quote from Skinner about how children first come to experience a sense of kinship with others, because in a sense, he does not think they ever do. Even adults are described as reacting solely to *behaviors* and not the hearts and minds of other people. Human beings have exquisite contingency detectors, and that is all there is to it. To use Skinner's phrase, human relations are largely a "matter of consequences," by which he means that people are only important to psychic development insofar as they "shape" the child's behavior by offering rewards and punishments as a consequence of the child's behavior.[5] In theory, a Skinner box would do just as well as a mother's loving embrace and adoring gaze if the contingencies were programmed correctly. There is nothing special about other people—it is an unscientific fiction that we feel a sense of connection to the moving mounds of meat we see out there.

Against these theories of scientists, there are philosophers (Smith, Hume, Husserl) and other thinkers who refused to portray the human from an isolationist perspective. Among them is René Girard and his Mimetic Theory. Girard proposed a theory of man that is anything but solipsistic, and instead insists on the importance of imitation and mimetic, or triangular, desire as a starting point.[6] He, along with Jean-Michel Oughourlian and Guy Lefort in *Things Hidden since the Foundation of the World* called for an *interdividual psychology*, based on self-other transactions rather than the isolated self.[7] In Girardian theory, imitation, mimetic desire, and other interpersonal transactions play a central role in the account of human affairs.[8]

In this chapter, I will show that if we are interested in how biology and culture cooperate to forge a person, if we are interested in the origins of self-other transactions, there is no better place to look than at developmental psychology. I will show that developmental psychology should be one discipline, among others, that influences, and is influenced by, Mimetic Theory.[9] Finally, I will show that developmental psychology, along with its sister discipline of neuroscience, can be a valuable member of a growing community of mimetic scholars and researchers as we jointly construct a new science of humanity.[10]

In this chapter, I discuss three areas of theory and empirical research that illuminate the science of infant intersubjectivity: imitation, gaze following, and infant intentionality. These three areas make manifest the rapport infants have with social others. The empirical discoveries in these areas provide a new and interesting foundation for Girard and Oughourlian's interdividual psychology.[11] I will consider broader theory in the last section of this essay, addressing a novel question: *Are babies natural-born Girardians?*

Sharing Others' Actions: Newborn Imitation

Background and Significance

Being caught up in others' movements and imitating what they are doing is an essential aspect of human sociality. Such imitation is in sharp contrast to the proclivities of monkeys and apes. Although other primates can learn in social settings, they do not duplicate the mannerisms used by another agent—instead, monkeys and apes simply pay attention to the endpoint and achieve it in whatever way suits them.[12] Similarly, the imitation of actions is deeply impaired in children with autism, and it has been theorized that the abnormality in imitation is one of the key factors that cuts them off from the social world.[13]

The imitation of body actions is sometimes dismissed as mere mimicry, and this has caused generations of psychologists to miss its centrality for humankind. On the contrary, action imitation is important for human rapport. In successful psychotherapy sessions, body mirroring is common between patient and therapist, and it is rampant between lovers and in parent-child relationships.[14] The duplication of the action patterns and gestures is part of the fabric of human communication; the imitation of customs, habits, rituals, practices, preferences, social norms, and values is essential to maintaining human culture itself.

Empirical Results and Implications

Meltzoff and Moore reported that 12- to 21-day-olds, and even newborn babes fresh from the womb, imitate facial acts.[15] The average age of the newborns was 36 hours old. The youngest child I tested was 42 minutes old, and she succeeded in imitating. Newborn facial imitation suggests an inborn mapping between the perception and production of human acts. We also discovered that neonates did not confuse either actions or body parts. Infants differentially responded to tongue protrusion with tongue protrusion and not lip protrusion, showing that a specific body part can be identified. Infants also differentially responded to lip protrusion versus lip opening, showing that differential action patterns can be imitated with the same body part. They even differentially imitated two separate kinds of acts with the tongue—one that is poking in-out from the midline, and the other that is poking in-out from the side of the mouth.[16] Following the 1977 discovery, there are now more than two dozen studies of early infant imitation from more than a dozen independent laboratories around the world.[17]

Newborns imitate, but this is not the end of the story. There is development in imitation. For example, newborns are certainly not self-conscious about imitating in the way older toddlers are, nor do they regulate imitation or use imitation functionally in the same way as older children and adults.[18] But growth and genesis is not incompatible with a rich starting state.

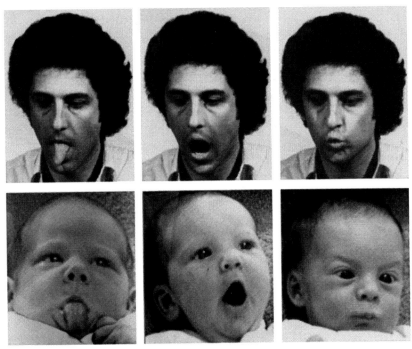

Figure 1. Developmental scientists discovered that human newborns imitate, suggesting that they are social beings at birth and can connect self and other, at least in some primitive fashion, even before language. This empirical discovery connects child psychology to issues in philosophy, anthropology, sociology, economics, and neuroscience.

Source: Andrew Meltzoff and Keith Moore, "Imitation of Facial and Manual Gestures by Human Neonates," *Science* 198 (1977): 75–78.

A Mechanism for Imitation: The AIM Hypothesis

Meltzoff and Moore proposed that early facial imitation is based on "active intermodal mapping" (AIM).[19] This is not a complex mechanism that requires deliberate cognitive machinations by the infant. The active nature of the matching process is undergirded by a proprioceptive feedback loop. The loop allows infants' motor performance to be compared against the seen target and allows infants to correct their response and home in on the target act. AIM proposes that comparison and correction is possible because the observation and execution of human acts are coded within a common framework. In our 1977 paper, we called it a *supramodal act space*, because it is not restricted to modality-specific information (visual, tactile, motor, etc.). Metaphorically, we said that exteroception (perception of others) and proprioception (perception of self) use the same code; there is no need for bringing the two together through prolonged learning and empiricist associations, because they are bound together at birth. A more detailed analysis of the infant's body scheme and the metric of equivalence between acts of self and other is provided elsewhere.[20]

This idea of supramodality that Meltzoff and Moore postulated is a psychological mechanism. It does not provide an analysis of the brain substrate. Recently, the supramodality we postulated 30 years ago based on infant imitation has garnered increased attention because of the neuroscience discovery of mirroring neural systems.[21] Such neural systems offer a subpersonal level of analysis, and developmental psychology provides a personal level of analysis—how the whole child perceives, interprets, and responds to a social other. Effort is being made to codify the commonalities and differences that derive from these different levels of analysis. The unique contribution from developmental psychology beyond the neuroscience work with animals and adult humans is that newborn imitation informs us about man's original nature—it demonstrates that self-other connectedness and communication exists at birth. *Humans imitate before they can use language; they learn through imitation but don't need to learn to imitate.*

OF SOCIAL TRIANGLES: SHARING OTHERS' ATTENTION

It is said that "The eyes are the window to the soul." This stirs the heart and mind of an empirical scientist. What does that mean? Can it be empirically studied? Is it true?

Background and Significance

The young infant lives in a kind of psychological Garden of Eden. There are two people paying attention to one another in a blissful state of dyadic inter-action. This does not last for long. Soon there are interlopers as the infant becomes aware that third parties are rivals for mother's affection. For example, infants begin to pay attention to the fact that mothers do not always look at them, but also cast their gaze on external objects, siblings, and spouses in the environment. One measure of this dawning realization is infants' gaze follow-ing—that is, their tendency to follow mother's gaze to an external target in order to see what she is looking at. Such gaze following is not the duplication of exact bodily movements, but rather a taking into account that mother's behavior is directed toward (or "about") an external target.

There is a triangle, to use a foundational Girardian concept. The triangle is formed by the mother–baby–object. In this triangle, the mother's visual glances refer infants to selected external targets. The onset of infant gaze following has profound implications both for language and emotions. It is relevant for understanding the meaning of an emotional display, because a person's emotion is often engendered by what he or she sees in the external world (e.g., that scene is disgusting or enticing). By following your partner's gaze, you can better understand the cause of her emotional display and thereby your partner's tastes, desires, and proclivities.

Language acquisition is also facilitated by following the gaze of social others (as strikingly described by St. Augustine in *The Confessions*, book 1). In the prototypical case, a mother's verbal label refers to the object she is look-ing at—*this* is a dog; *that* is a bottle—where the referent is indicated by the mother's gaze. We have empirically demonstrated that individual infants who follow mother's gaze move more quickly through the first stages of language learning.[22] Beyond all this, some, like Ginger Rogers, would argue that gaze following is an intersubjective act: "When two people love each other, they don't look at each other, they look in the same direction."

In developmental psychology, there is a debate about the mechanism underlying gaze following and its developmental time course. One deflation-ary proposal suggests that following where another looks has nothing to do with intersubjective sharing. According to this point of view, the behavior is based on infants' visually tracking the adult's head movements because the head produces salient/large displacements in the visual field. Inasmuch as infants visually follow the movements of this physical object in space (the head), they are automatically "dragged" to the correct half of space. Once in the correct hemi-field, they latch onto whatever attractive object is there, usu-ally the same one at which the adult is looking. In this instance, infants are *not* responding intersubjectively; they are simply processing physical movements in space caused by the head. They would be just as likely to follow the move-ments of an inanimate object.

The alternative, richer view is that infants are turning to look at what the adult is seeing. It is an important step forward in intersubjective understand-ing to put special emphasis on human gaze. It is the eyes that are the "win-dow to the soul"—the head is not such a portal. Proust, one of the primary novelists used by Girard to develop his concept of triangular desire, knew the importance of eyes: "It's such a nuisance to think that whatever insignificant thing you may be doing, other eyes are watching you." "That smile fell on me, who had not taken my eyes off of her. Recalling, then, the gaze she had rested on me during the Mass, as blue as a ray of sunlight passing through Gilbert the Bad's window, I said to myself: 'Why she's actually paying attention to me.' I believed that she liked me, that she would still be thinking of me after she had left the church, that because of me perhaps she would be sad that evening at Guermantes. And immediately I loved her."[23]

Empirical Results and Implications

We conducted an empirical study on the development of gaze following in children.[24] An adult turned to look at one of two targets. The adult turned to the target with eyes open for one group and with eyes closed for the other group. If infants relied simply on head motions, they should turn in both cases, because head movement was controlled. If, however, infants appreciate that the eyes are relevant for connecting a "perceiver" and object, then they should turn to look at the target in one situation and not the other.

The children were all between 1 and 1.5 years old at the time of the tests. We found that infants followed the adult significantly more often when the adult turned with open versus closed eyes. The nonsocial interpretation of gaze following is that a head movement simply drags infants' attention to a hemi-field of space. The current findings disprove this interpretation, because the head movement was the same in both conditions, yet the infants followed the adult to the external object in one case (eyes open) and not the other. In line with Sartre's thoughts about the importance of gaze in bringing psychological attention to people and things, and Girard's further observation that such a gaze may "transfigure" the object or "confer upon it illusory value," we discovered that an inanimate object takes on a special valence when it is looked at by a social other.[25] It is as if having the adult shine her social spotlight on an inanimate object leaves a trace on it, an invisible mark. Such is the power of eyes, that being visually touched by the look of a social other transforms the object from a boring blob to an object of desire that cries out, "Look at me! Value me!"

Figure 2. A primordial triangle. Gaze following brings infants into perceptual contact with objects that are being attended to by adults. This provides perceptual conditions for infants' catching adult desires.

Source: Andrew Meltzoff et al., "Foundations for a New Science of Learning," *Science* 325 (2009): 284–88.

Eye closure is only one way to block a person's line of sight. Another way is to use an inanimate object. From an adult perspective, an opaque physical barrier and closed eyes both have similar effects on human perception—both prevent visual access. But infants may interpret the actions of people (eye closures) differently from the positioning of inanimate objects. They may give different meanings to the physical events than adults do.

In our study of inanimate occluders, the person turned toward a target wearing either a headband or a blindfold.[26] We discovered that 14- and 18-month-old infants looked at the adult's target significantly more often in the headband than the blindfold condition. In contrast, the 12-month-old infants looked equal amounts in both conditions. It is not that their looking was suppressed, but rather that they looked at the specified object despite the fact that the adult had her eyes covered by the blindfold when she turned to "look" at it.

Recall that the 12-month-olds succeeded on the eyes closed/open test, but when the adult's vision was blocked by an inanimate object (blindfold), they failed. It is as if they recognize that the human act of eye closure blocks contact with external objects, but do not yet understand the same about inanimate occluders. How could this be? What psychological mechanisms are at play?

"Like Me" and "Like You": Shared Experiences

Background and Significance

The foregoing research suggests that infants understand that eye closure blocks perception earlier than they understand that inanimate objects block perception. My hypothesis is that this is because infants themselves have subjective experience with the perceptual effects of their own eye closures. When they close their eyes, the world goes black. They have control over this, and massive agentive experience. I hypothesize that they use such experience to imbue the eye closures of others with felt meaning.

This hypothesis leads to a novel prediction. Nonbiological occluders should become more meaningful if infants are themselves given opportunities to learn that they block their own vision.

Empirical Results and Implications

Meltzoff and Brooks gave 12-month-olds experience that blindfolds lead to psychological effects—that the infants themselves cannot see through an inanimate barrier.[27] In order to provide them with this experience, a set of desirable objects were put out on the table, and whenever the infant looked at one, the adult would bring the blindfold up and block the child's line of sight. Their perception of the world was impeded when the blindfold was held in front of *their* eyes, and was restored again when the blindfold was removed. This subjective experience had nothing to do with the experimenter's viewpoint; it was a first-person experience.

In the critical test, the adult put the blindfold over her own eyes. This was the first time the infants were presented with the blindfolded adult. The blindfolded adult then turned toward an external object just as in the normal gaze-following test. The question was whether the infant would follow the adult. In order to rigorously assess the effects of this blindfold training, we used two control groups. In one control group, infants simply experienced baseline familiarization with the blindfold (they played with the black cloth while it lay flat on the table and did not experience that it blocked their own view). In a second control group, we cut a small "peeking window" out of the middle of the cloth so that infants could see through it whenever it was put in front of their eyes; it did not block their view of the desired object when the adult brought it up in front of their eyes.

We found that the nature of infants' self-experience with the blindfold completely changed their interpretation of the actions of the blindfolded other. Infants who had self-experience that the blindfold blocked their own view now appreciated the consequences of blindfolds for the other: They did not turn when the adult wore the blindfold. In contrast, as we had predicted, in both of the control groups (baseline and cloth with peeking window) the infants still mistakenly followed the blindfolded adult's "gaze."

This experimental demonstration shows that infants' first-person experience transforms their interpretation of the subjective experiences of others. These effects provide a confirmation of Meltzoff's "Like Me" framework of the growth of social understanding.[28] Although I call it the "Like Me" framework, I would also like to emphasize the bidirectionality of the self-other relationship; the longer name I prefer to use is the "Like Me–Like You" Framework. This idea fundamentally revises the solipsism of Piaget, and lays the groundwork for a new view of infant development within the context of intersubjectivity. The more elaborated ontogenetic and philosophical implications of this developmental framework are elaborated in the conclusions of this essay and elsewhere.[29]

SHARING OTHERS' GOALS AND INTENTIONS: WHAT IS A "PERSON" TO AN INFANT?

Background and Significance

In the mature adult system of understanding and emotions, I not only share behavioral actions and visual perceptions with others, but I also share in their goals and underlying intentions. Intentions are particularly interesting, because they lie deeper than the plane of action and are related to desires, a most Girardian concept.

A first question is whether infants have any inkling of the distinction between the actions someone performs and their intention in performing these actions. Wittgenstein clarifies this distinction with this pithy insight: "What is left over if I subtract the fact that my arm goes up from the fact that I raise my arm?" Answer: "Intention."[30]

Is there any evidence that infants have a feel for human acts that penetrates below the surface behaviors to the intentions that lie behind them?

Empirical Results and Implications

The "behavioral re-enactment procedure" was created to investigate infants' reactions to the goals and intentions of others.[31] The procedure capitalizes on children's tendency to reenact or imitate, but uses it in a more abstract way to investigate whether infants can read below the literal surface behavior to something like the goal or intention of the actor. The procedure involves showing infants an unsuccessful act. For example, the adult accidentally under- or overshoots his target—or he tries to pull apart a dumbbell-shaped toy, but his hand slips off the ends and he is unsuccessful. Thus the goal state is not achieved. Adults immediately sense the actor's intentions, although he never fulfills them. The question is whether children see beyond the literal body movements to the goal or intention of the act. In a sense, the "correct answer" is to not copy the literal movement, but the intended act that remains unfulfilled and invisible.

In my 1995 paper on infant intentionality, I showed 18-month-old infants an unsuccessful act. The study compared infants' tendency to perform the target act in several situations: after they saw the full target act demonstrated, after they saw the unsuccessful attempt to perform the act, and after it was neither shown nor attempted.

I found that 18-month-olds can infer the unseen goals implied by unsuccessful attempts. Infants who saw the unsuccessful attempt and infants who saw the full target act both produced target acts at a significantly higher rate than controls. Infants seemed to "see through" the surface behavior to the underlying goals or intentions of the actor. Evidently, toddlers can understand our goals even if we fail to fulfill them.

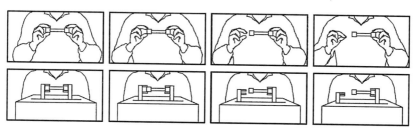

Figure 3. Infants treat people differently from things. Human demonstrator (*top panel*) and inanimate device tracing the same spatiotemporal movements (*bottom panel*). Infants treat the human acts as indicating an intention to pull the object apart; they do not interpret the movements of the inanimate device in this way.

Source: Andrew Meltzoff, "Understanding the Intentions of Others: Re-enactment of Intended Acts by Eighteen-Month-Old Children," *Developmental Psychology* 31 (1995): 838–50.

Psychology vs. Physics: The Goals of People and the Motions of Machines

In the adult framework, only certain types of entities are ascribed intention and purposiveness. Chairs and boulders rock and roll, but their motions are not seen as intentional. Most prototypically, human acts are the types of movement patterns that are seen as caused by intentions. What do infants think?

To begin to examine this, I tested how 18-month-olds responded to a mechanical device that mimicked the same movements as the actor in the unsuccessful-attempt condition.[32] An inanimate device was constructed that had poles for arms and mechanical pincers for hands. It did not look human, but it could move very similarly to the human. For the test, the pincers "grasped" the dumbbell at the two ends just as the human hands did. One mechanical arm was then moved outwards, just as in the human case, and its pincer slipped off the end of the dumbbell just as the human hand did. The movement patterns of machine and man were closely matched from a purely spatiotemporal description of movements in space.

I found that infants did not attribute a goal or intention to the movements of the inanimate device. Although they were not frightened by the device and looked at it as long as at the human display, they simply did not see the sequence of actions as implying a goal. All infants picked up and explored the toy, but they were no more (or less) likely to pull apart the toy after seeing the manipulation by the inanimate device as in baseline conditions when they saw nothing.

I think 18-month-olds cast the person's actions within an intersubjective framework that differentiates between the visible behavior and a deeper level of felt experience involving human goals and intentions. When they watch a person's hands slip off the ends of the dumbbell, they immediately see what the adult was "trying" or "striving" to do. When they see the inanimate device slip off the end of the dumbbell, they see it as mere mechanical slippage with no intentionality.[33]

INTERSUBJECTIVITY BEFORE LANGUAGE: A RAPPROCHEMENT BETWEEN SCIENCE AND THE HUMANITIES

The Problem

According to classical theories of psychology, from Freud to Piaget, newborns lack any sense that humans are "subjects" different from the "objects" in the environment. It was claimed, for example, that the child is born a solipsist or is in a state of "normal autism," treating people the same as things.[34] Some philosophers have also suggested this and have conjectured about what the original state of human beings "must be" based on the logic from one school of thought or another.

In the modern era, the science of developmental psychology has taught us new things. First, it has taught us to take seriously the question of genesis: "How can we get here from there?" If we are born solipsistic or autistic, what sort of experiences would subsequently lead us to identify with our fellow humans, to feel empathic pain, and to be driven to communicate, cooperate, and compete with them? No one has been able to provide an adequate account of this transformation. This is not a good sign.

Second, a kind of "natural experiment" has been done. There is no evidence that either home-reared chimpanzees, or children with autism develop the same "fellow feelings" or sense of intersubjectivity felt by typical adults. Without a certain biological substrate, intersubjectivity cannot be created by cultural immersion alone.

Taken together, this means that building an adequate science of humanity requires a theory that incorporates both structure and genesis. To paraphrase Kant, theories of innateness that deny genesis have no future; theories of genesis that deny initial biological structure are groundless.

Intersubjectivity as the Starting Point for Human Development

Philosophers' queries about man's original nature cannot be answered by tests of adults (whether monkey or man). We need tests of infants. Newborn imitation indicates that at some level of processing, no matter how primitive, infants can map acts of other people onto acts of their own body. Because human acts are seen in others and performed by the self, the infant can grasp the bidirectional interpersonal connection: You can act "like me" and I can act "like you," which is the basis of my proposed "Like Me–Like You" developmental framework.[35] This self-other equivalence provides preverbal children a privileged access to people not afforded by things, and a basis for sharing and communication.

It has long been thought that the equivalence between self and other is important for normal adult psychology, and classical philosophers such as Adam Smith and David Hume have explored how it leads to "fellow feelings" for others, as described by Jean-Pierre Dupuy.[36] Human empathy, pity, and putting oneself in someone else's shoes to understand their point of view seem to depend on this. But there is a lingering dilemma: The problem has always been that this equivalence was thought to be a late achievement, possibly dependent on language.

The findings from developmental science suggest that preverbal human infants immediately register similarities between self and other. This is not a derived, complex, or cognitively advanced analysis of the social world. There is an intrinsic identification that infants feel before they can speak. This felt connection colors infants' first interactions and interpretations of the social world and impels human communication and social development. Other people are viewed as "Like Me" from the start. This is the newborn's first glimpse of humanity. It is not derived, but basic—not the culmination of moral sentiments, but the foundation of them.

Normal adults feel a "Like Me" kinship with their own children and family; moral and religious leaders have tried to teach us to widen this circle and feel kinship with all people regardless of their station.[37] But babies—so most scientific, philosophical, and religious theories say—do not intrinsically sense this connection. Children are taught to feel this way. They begin life as asocial, solipsistic creatures with no sense of the "Like-me-ness" of others, no bridge to the souls of others. Starting from this impoverished state, they are then "socialized." The current research in developmental science shows that this is incorrect. Infants, even newborns, see others as being "Like Me." This basic connectedness is the progenitor of a subsequent, more elaborate grasp of self-other relatedness.

Embodiment and a Mechanism for Change in Intersubjectivity

If infants begin life in this rich state, how does further development occur? I believe that the "Like Me" perception coupled with social interaction provides them with an engine for developmental change and transformation. For example, the infant knows that when she desires something, she reaches out and grasps it. The infant experiences her own internal desires and the concomitant bodily movements (hand extension, finger movements, etc.). The experience of grasping to satisfy desires gives infants leverage for "feeling with" the other who grasps for things. When the child sees another person reaching for an object, she sees the person extending his hand in the same way. These movements are imbued with experiential meaning, in part because of the child's own past experience with these acts. This adds, I think, a new dimension to Girard's idea about mimetic desire.

A similar argument applies to the goal-directed "striving" and "try and try again" behavior in my studies using the behavioral reenactment procedure. Infants have experienced their own thwarted desires, failed plans, and unfulfilled intentions. This *intra*subjective experience deepens their *inter*subjective grasp of the motivation and meaning of others' behavior. When an infant sees another act in this same way, the infant's self-experience could suggest that there is a purpose, desire, or intention beyond the visible bodily movements that are present to the senses. Infants would see the adult's unsuccessful attempts, and the behavioral envelope in which they occur, as a pattern of strivings rather than ends in themselves.

Gaze following admits to a similar theoretical analysis. The understanding of another's looking behavior could benefit from *intra*subjective experiences—in this case, experience of oneself as a perceiver. Infants' experience is that eye closure cuts off their own perceptual access. This experience provides a basis for imbuing the eye closures of others with felt meaning.

The Future

A stumbling block for classical theories of human psychology was that the self-other equivalence was postulated to be late developing—emerging from language or deliberate and complex cognitive analyses. My research stands this proposition on its head. It suggests that young infants register the acts of others and their own acts in commensurate terms. The recognition of self-other equivalences is the starting point for reacting to other humans—a precondition for human development, not the outcome of it.

Is the Human Infant a Natural Girardian?

My research suggests that infants may be "natural Girardians," at least with regard to two principle philosophical claims in his writing. First, imitation operates prior to language and other formal symbolic skills, and in turn is a primary precondition for their genesis; second, imitation extends beyond mere surface gestures and behaviors to underlying goals, desires, or intentions. In contrast to previous generations of scientific thought, Girard's mimetic theory hypothesizes a primordial role for imitation in the genesis of human culture, cognition, and sociality. From a developmental perspective, my research corroborates and elaborates this claim by demonstrating that infants are able to imitate much earlier in life than we previously suspected, and that they subsequently follow the gaze of others in a way that draws them into the powerful orbit of adult behavior, goals, intentions, and desires. Infant imitation serves as the starting state that supports learning and the genesis of new and dynamic forms of human cognition and sociality based on interpersonal interaction, including mutual-informing imitation between self and other.[38]

Are Children Natural-Born Girardians with Respect to Mimetic Rivalry?

In addition to addressing "positive" or "cooperative" learning and sociality, Girard's mimetic theory, as articulated by several authors in this volume and others, also makes predictions concerning the role of imitation in the emergence of human desire, rivalry, envy, and violence.[39] Girard's observations seem to be particularly relevant to a variety of phenomena that can be observed in children, including sibling rivalry for inanimate things (which can begin early and involve competition over common objects and toys) as well as rivalry for parental attention (which itself builds on the earlier sensitivity to gaze direction and attention, but now applies to self and other).[40] I have begun some research examining mimetic desire and the roots of rivalry, but the work is in the initial phases.[41] Clearly, there are important aspects of Girardian theory that are inadequately explored from an empirical viewpoint, not only regarding violence but also regarding the developmental pathway from joint desire to rivalry and even to bullying or scapegoating in a crowd of children.

What is now to be done? There should be no rivalry between armchair philosophy and laboratory research in the quest for constructing a science of humanity that incorporates a developmental perspective—one that adequately describes man's original nature and how we get ourselves from *there* (thinking, feeling newborns) to *here* (philosophical adults). The messages from the mouths of babes are too important to ignore, and too complex to understand without multidisciplinary collaboration. To adapt Girard: Babies hold a secret about the human mind that has been hidden for millennia. They are our double. They have a primordial drive to understand us that advances their development; we have a desire to understand them that propels social science and philosophy. By examining the minds and hearts of children, we illuminate ourselves.

ACKNOWLEDGMENTS

I am especially grateful to Scott Garrels for his kind patience and intellectual assistance as editor of this volume. I thank Eugene Webb for introducing me to Girard's thinking and for providing comments on this essay. I am deeply indebted to René Girard for insightful conversations about imitation and human nature, and to the other authors in this volume for providing comments on earlier versions of this essay during provocative meetings in Paris, Innsbruck, and Stanford. I thank the National Institutes of Health (HD-22514), the National Science Foundation (OMA-0835854), and the Tamaki Foundation for generous financial support.

NOTES

1. Sigmund Freud, "Formulations on the Two Principles of Mental Functioning," in *The Standard Edition of the Complete Psychological Works of Sigmund Freud*, ed. J. Strachey (London: Hogarth Press, 1911), 12:220.

2. Jean Piaget, *The Origins of Intelligence in Children*, trans. M. Cook (New York: International Universities Press, 1952).

3. Jean Piaget, *The Construction of Reality in the Child*, trans. M. Cook (New York: Basic Books, 1954).

4. Burrhus Skinner, *Science and Human Behavior* (New York: Macmillan, 1953).

5. Burrhus Skinner, *A Matter of Consequences* (New York: Alfred A. Knopf, 1983).

6. René Girard, *Deceit, Desire, and the Novel: Self and Other in Literary Structure*, trans. Yvonne Freccero (Baltimore: Johns Hopkins University Press, 1965).

7. René Girard, *Things Hidden since the Foundation of the World*, trans. S. Bann and M. Metteer (Stanford, CA: Stanford University Press, 1987).

8. Eugene Webb, *The Self Between: From Freud to the New Social Psychology of France* (Seattle: University of Washington Press, 1993). See also Jean-Michel Oughourlian, "From Universal Mimesis to the Self Formed by Desire," this volume.

9. Jean-Pierre Dupuy, "Intersubjectivity and Embodiment," *Journal of Bioeconomics* 6 (2004): 275–94. See also Jean-Pierre Dupuy, "Naturalizing Mimetic Theory," this volume.

10. Vittorio Gallese, "The Manifold Nature of Interpersonal Relations: The Quest for a Common Mechanism," *Philosophical Transactions of the Royal Society of London, Series B, Biological Sciences* 358 (2003): 517–28; Vittorio Gallese, "Before and Below 'Theory of Mind': Embodied Simulation and the Neural Correlates of Social Cognition," *Philosophical Transactions of the Royal Society of London, Series B, Biological Sciences* 362 (2007): 659–69; Andrew Meltzoff and Jean Decety, "What Imitation Tells Us about Social Cognition: A Rapprochement between Developmental Psychology and Cognitive Neuroscience," *Philosophical Transactions of the Royal Society of London, Series B, Biological Sciences* 358 (2003): 491–500; Philip Jackson et al., "How Do We Perceive the Pain of Others? A Window into the Neural Processes Involved in Empathy," *NeuroImage* 24 (2005): 771–79.

See Vittorio Gallese, "The Two Sides of Mimesis: Mimetic Theory, Embodied Simulation, and Social Identification," this volume.

11. Jean-Michel Oughourlian, *The Puppet of Desire: The Psychology of Hysteria, Possession, and Hypnosis*, trans. with an intro. by Eugene Webb (Stanford, CA: Stanford University Press, 1991); Jean-Michel Oughourlian, *The Genesis of Desire*, trans. Eugene Webb (East Lansing: Michigan State University Press, 2010). See Jean-Michel Oughourlian, "From Universal Mimesis to the Self Formed by Desire," this volume.

12. Andrew Meltzoff, "Imitation, Objects, Tools, and the Rudiments of Language in Human Ontogeny," *Human Evolution* 3 (1988): 45–64; Michael Tomasello et al., "Cultural Learning," *Behavioral and Brain Sciences* 16 (1993): 495–552.

13. Geraldine Dawson et al., "Neuropsychological Correlates of Early Symptoms of Autism," *Child Development* 69 (1998): 1276–85; R. Peter Hobson and Anthony Lee, "Imitation and Identification in Autism," *Journal of Child Psychology and Psychiatry* 40 (1999): 649–59; Karen Toth et al., "Early Predictors of Communication Development in Young Children with Autism Spectrum Disorder: Joint Attention, Imitation, and Toy Play," *Journal of Autism and Developmental Disorders* 36 (2006): 993–1005. See Ann Cale Kruger, "Imitation, Communion, and Culture," this volume.

14. Scott Garrels, "Imitation, Mirror Neurons, and Mimetic Desire: Convergence between the Mimetic Theory of René Girard and Empirical Research on Imitation," *Contagion: Journal of Violence, Mimesis, and Culture* 12–13 (2006): 47–86; Vittorio Gallese et al., "Intentional Attunement: Mirror Neurons and the Neural Underpinnings of Interpersonal Relations," *Journal of the American Psychoanalytic Association* 55 (2007): 131–76; Jean-Michel Oughourlian, *The Genesis of Desire.*

15. Andrew Meltzoff and Keith Moore, "Imitation of Facial and Manual Gestures by Human Neonates," *Science* 198 (1977): 75–78; Andrew Meltzoff and Keith Moore, "Newborn Infants Imitate Adult Facial Gestures," *Child Development* 54 (1983): 702–9; Andrew Meltzoff and Keith Moore, "Imitation in Newborn Infants: Exploring the Range of Gestures Imitated and the Underlying Mechanisms," *Developmental Psychology* 25 (1989): 954–62.

16. Andrew Meltzoff and Keith Moore," Imitation, Memory, and the Representation of Persons," *Infant Behavior and Development* 17 (1994): 83–99.

17. For a review, see Andrew Meltzoff and Keith Moore, "Explaining Facial Imitation: A Theoretical Model," *Early Development and Parenting* 6 (1997): 179–92.

18. Betty Repacholi and Andrew Meltzoff, "Emotional Eavesdropping: Infants Selectively Respond to Indirect Emotional Signals," *Child Development* 78 (2007): 503–21; Rebecca Williamson et al., "Learning the Rules: Observation and Imitation of a Sorting Strategy by 36-Month-Old Children," *Developmental Psychology* 46 (2010): 57–65.

19. Meltzoff and Moore, *Explaining Facial Imitation,* 179–92.

20. Ibid.

21. Vittorio Gallese et al., "Action Recognition in the Premotor Cortex," *Brain* 119 (1996): 593–609; Vittorio Gallese, *Manifold Interpersonal Relations,* 517–28; Giacomo Rizzolatti et al., "Premotor Cortex and the Recognition of Motor Actions," *Cognitive Brain Research* 3 (1996): 131–41; Giacomo Rizzolatti and Laila Craighero, "The Mirror-Neuron System," *Annual Review of Neuroscience* 27 (2004): 169–92.

22. Rechele Brooks and Andrew Meltzoff, "Infant Gaze Following and Pointing Predict Accelerated Vocabulary Growth through Two Years of Age: A Longitudinal, Growth Curve Modeling Study," *Journal of Child Language* 35 (2008): 207–20; M. Carpenter et al., "Social Cognition, Joint Attention, and Communicative Competence from 9 to 15 Months of Age," *Monographs of the Society for Research in Child Development* 63 (1998).

23. Marcel Proust, *Swann's Way,* trans. L. Davis (New York: Penguin, 2002), 164, 181.

24. Rechele Brooks and Andrew Meltzoff, "The Importance of Eyes: How Infants Interpret Adult Looking Behavior," *Developmental Psychology* 38 (2002): 958–66.

25. Jean-Paul Sartre, *Being and Nothingness*, trans. H. E. Barnes (London: Methuen, 1957); Girard, *Deceit, Desire, and the Novel*, 17.

26. Meltzoff and Brooks, *Importance of Eyes*, 958–66.

27. Andrew Meltzoff and Rechele Brooks, "Self-Experience as a Mechanism for Learning about Others: A Training Study in Social Cognition," *Developmental Psychology* 44 (2008): 1257–65.

28. Andrew Meltzoff, "'Like Me': A Foundation for Social Cognition," *Developmental Science* 10 (2007): 126–34.

29. Andrew Meltzoff, "The 'Like Me' Framework for Recognizing and Becoming an Intentional Agent," *Acta Psychologica* 124 (2007): 26–43; Andrew Meltzoff et al., "Foundations for a New Science of Learning," *Science* 325 (2009): 284–88.

30. Ludwig Wittgenstein, *Philosophical Investigations*, trans. G. E. M. Anscombe (Oxford: Blackwell, 1953).

31. Andrew Meltzoff, "Understanding the Intentions of Others: Re-enactment of Intended Acts by 18-Month-Old Children," *Developmental Psychology* 31 (1995): 838–50.

32. Ibid.

33. It is possible that displays can be constructed that fool infants, as they do adults. Can a robot be considered intentional? We do not know the full set of conditions necessary for infants seeing intentionality in the "other." Andrew Meltzoff is working on infant robotics; Vittorio Gallese ("The Two Sides of Mimesis"), Paul Dumouchel ("Emotions and Mimesis"), and others in this volume are interested in human reaction to robots for similar reasons. For developmental robotics, see Aaron Shon et al., "A Cognitive Model of Imitative Development in Humans and Machines," *International Journal of Humanoid Robotics* 4 (2007): 387–406, and Andrew Meltzoff et al., "Social Robots are Psychological Agents for Infants: A Test of Gaze Following," *Neural Networks* 23 (2011): 966–72.

34. See Piaget, *The Construction of Reality in the Child*; and Margaret Mahler et al., *The Psychological Birth of the Human Infant* (New York: Basic Books, 1975).

35. Andrew Meltzoff, *Foundation for Social Cognition*, 126–34; Andrew Meltzoff, *Framework for Recognizing and Becoming an Intentional Agent*, 26–43; Andrew Meltzoff and Rechele Brooks, *Self-Experience as Mechanism*, 1257–65.

36. Jean-Pierre Dupuy, *Intersubjectivity and Embodiment*, 275–94. See also Jean-Pierre Dupuy, "Naturalizing Mimetic Theory," this volume.

37. Eugene Webb, *Worldview and Mind: Religious Thought and Psychological Development* (Columbia: University of Missouri Press, 2009). For neuroscience work on empathy for others who are different from oneself, see Claus Lamm et al., "How Do We Empathize with Someone Who Is Not Like Us? A Functional Magnetic Resonance Imaging Study," *Journal of Cognitive Neuroscience* 22 (2010): 362–76.

38. See Meltzoff, *Foundation for Social Cognition*, 126–34, for an empirical and theoretical examination of mutual imitation; Jean-Pierre Dupuy, "Naturalizing Mimetic Theory," in this volume, for the morphogenetic function of imitation; and Meltzoff et al., *New Science of Learning*, 284–88, for how imitation and sociality influence language acquisition, learning, and education.

39. Eugene Webb, *Worldview and Mind*, 2009; Robert Hamerton-Kelly, *Politics and Apocalypse* (East Lansing: Michigan State University Press, 2008). See also Jean-Michel Oughourlian,

"From Universal Mimesis to the Self Formed by Desire," Paul Dumouchel, "Emotions and Mimesis," Mark R. Anspach, "Imitation and Violence: Empirical Evidence and the Mimetic Model," and Jean-Pierre Dupuy, "Naturalizing Mimetic Theory," all in this volume.

40. Betty Repacholi et al., "Infants' Understanding of the Link between Visual Perception and Emotion: 'If She Can't See Me Doing It, She Won't Get Angry,'" *Developmental Psychology* 44 (2008): 561–74.

41. Cristina Atance and Andrew Meltzoff, "Preschoolers' Current Desires Warp Their Choices for the Future," *Psychological Science* 17 (2006): 583–87; Cristina Atance et al., "Preschoolers' Understanding of Others' Desires: Fulfilling Mine Enhances My Understanding of Yours," *Developmental Psychology* (2011); Yawei Cheng et al., "Motivation Modulates the Activity of the Human Mirror-Neuron System," *Cerebral Cortex* 17 (2007): 1979–86.

CHAPTER 4

Emotions and Mimesis

Paul Dumouchel

When two young men grow up together, they learn the same lessons, read the same books, play the same games, and agree on just about everything. They also tend to desire the same objects. This perpetual convergence is not incidental, but essential to their friendship; it occurs so regularly and inevitably that it seems preordained by some supernatural fate; it really depends on a mutual *imitation* so spontaneous and constant that it remains unconscious.

Eros cannot be shared in the same manner as a book, a bottle of wine, a piece of music, a beautiful landscape. Proteus is still doing what he has always done—imitating his friend—but this time the consequences are radically different. All of a sudden, with no advance warning, the attitude that has always nourished the friendship tears it apart. Thus imitation is a double-edged sword. At times it produces so much harmony that it can pass for the blandest and dullest of all human drives; at other times it produces so much strife that we refuse to recognize it as imitation.[1]

Thus writes René Girard towards the beginning of *A Theater of Envy: William Shakespeare*, when he presents Proteus and Valentine, the two gentlemen of Verona. The text quoted above assumes a strange and complex relationship between emotions and imitation. The first paragraph suggests that a spontaneous, constant, and unconscious imitation is not only an indispensable part of the two young men's friendship but also, to some extent, the process through which it came to be established. It suggests that mimesis is part of the process through which some emotions are stabilized. The second paragraph, or rather the incident to which it refers (Valentine, seeking the approval of Proteus, presents to him and praises his love, Sylvia, and Proteus, inevitably imitating Valentine in this as he does in everything else, falls in love with Sylvia and becomes his friend's rival) suggests a somewhat different relation between the two phenomena. Here it is apparently the friendship between the two young men that explains the imitation. Proteus is all the more sensitive to his friend's suggestion because he is his friend. However, there is a rhetorical context here, and Girard should not be understood as suggesting that if the two men had not been friends, rivalry could not have occurred; rather his point is that conflict emerges through the very same process that made them friends.[2] Because the two men are friends, their sudden conflict and Proteus's treason seem incomprehensible. We fail to recognize that these changes in Proteus's behavior constitute but a further form of imitation, and thus at another, more abstract level of description constitute the continuation of the same behavior, and like Valentine, we do not understand that the "victim" has helped to bring about the evil that befalls him. Both paragraphs taken together suggest that emotions and mimesis are so intertwined that it may well be impossible to separate them and to determine the particular contribution of each in human behavior.

Public Emotions and Private Imitation

However, there is at least one thing that is quite different between the two phenomena that may constitute a good starting point for this analysis. Emotions are public and evident in at least two senses. Proteus and Valentine have been friends since they were children—that they are close to each other is a thing rapidly evident to all, and so will it be with their subsequent rivalry. Emotions, of course, may be hidden, and people can lie about their intentions towards each other, but the very fact that they do proves that emotions are public and evident cues to agents' future actions. Without this prima facie publicity of emotions, there would never arise any need to deceive and dissimulate one's true feelings. Emotions are also evident in another sense: they are the normal denizens of folk psychology. Mimesis is not. That Proteus and Valentine actually imitate each other, both when they are friends and when they are enemies, is not something that is evident. It takes all the genius of Shakespeare, and that of Girard, to render visible this type of relation between agents that determines their actions, but it is for the most part unconscious, and they are usually unaware of it. To the contrary, emotions are part of the normal stuff of everyday life. It may be that we do not understand very well how emotions function and the role they play in our life; nonetheless they are ordinary phenomena that function as part of our explanation of people's behavior, whether it is in the context of everyday life or in that of scientific psychology.

That difference is important. Why is it that a phenomenon as fundamental as mimesis—if we are to believe Girard, all of human culture ultimately stems from it—has so rarely been recognized as such? While emotions are, so to speak, everywhere and for anyone to see, mimesis as a whole has remained hidden "since the foundation of the world." This suggests that the relationship between emotions and overt behavior is much more transparent than is the relation of such behavior with mimesis.[3] It is true that the absence of imitation in our folk psychology may to some extent be a curiosity proper to modern Western culture. Many cultures put much more stress on the importance and value of imitation as a means of learning, and on the beneficial effects of imitation in general.[4] However, few cultures have had more than a passing intuition concerning the continuity between imitation and conflict, which, one can argue, is precisely what defines *mimesis* as opposed to imitation in a more traditional sense. So what is it that explains this difference?

One answer that is rather evident, which I suspect is important, is that *stricto sensu* there is no experience of mimesis, and very little of imitation. You may surprise yourself imitating your hated rival or your friend, but when that happens, which I think is rare, what you experience is perhaps surprise at this discovery, but not imitation itself, which does not feel like anything. Unlike emotions, mimesis is not an object of direct first-person perception. Mimesis is something that mainly appears from the third-person point of view, while emotions are primarily given to us from the first- and second-person points of view. That is to say, we discover emotions either as subjective experiences, for example the feeling of being angry or happy, or by being at the receiving end of another person's anger or happiness. In fact, even when they are viewed from a third person's point of view, emotions, interestingly enough, also give rise to a specific emotional experience, which has classically been described as that of empathy or sympathy—an important topic to which we will return later on. On the other hand, the "experience" of imitating, for example, as a result of explicit instruction, as when you are learning a sport or craft, typically disappears into the experience of doing whatever it is that you are doing, archery or baking a cake, and the particular feeling of succeeding or failing at it.

Current research on postural mimicry and low-level imitation tends to confirm this. There is now a lot of evidence of how humans and other primates tend to copy each other's behavior and adopt the same body position when interacting. Agents are usually not aware of this, and when they suddenly realize that they are folding their arms or crossing their legs in the same way as the person with whom they are conversing, they usually break the symmetry immediately. This change of behavior on the part of the imitating agent when he becomes aware of his imitation may be related to the fact that explicit imitation among humans is related either to communal agreement (as in song, dances, and many rituals), or to exclusion and ostracism (as in mockery, which often takes the form of explicit imitation of the victim's ticks or physical defects). In both cases, imitation is a powerful social signal, and many monkeys seem clearly aware of when they are being imitated.[5] This suggests that there is no first-person direct experience of imitation, but only a kind of third-person-level perception of imitating: one does not experience imitating, but sees (surprises) oneself imitating as from the outside, through the eyes of the other. Furthermore, the change of behavior to which this discovery leads seems related to the experience of *being imitated*, which is immediately perceived as charged with potent signification. Of imitation there is little and generally only passing experience, and of mimesis there is none.

In consequence, Shakespeare can speak directly of his characters' emotions; he can let them express their feelings, intentions, and desires as they experience them. However, he can only reveal mimesis and imitation indirectly through their actions, through a mise en scène that makes visible for the third party (audience or reader) the symmetry of their behavior. At best, he can include in the play a third party who observes the actions of the principals and indicates to the spectators where to look. Emotions, on the other hand, tend to be excessively present. Emotions occupy the front of the stage. The poet's problem with emotions is not to find a way to make them visible, but to tone down their excessive presence, to abate their outrageous evidence as explanations of actions. Of course, declarations of love, expressions of anger, promises of fidelity should not always be taken at face value; however, it is an important part of what they are that they have that face value. Otherwise the story would have no sense.

Unlike love, power politics, jealousy, or revenge, mimesis is never, and can never be, the explicit theme of a play. In other words, mimesis makes its presence and influence felt through other things—jealousy, honor, deception, love, or treason. These things, among them emotions, give themselves to us directly. The problem then is that mimesis never gives itself to us directly. It can be observed, but only one level removed from where it is taking place.

Why then is mimesis invisible while emotions are, on the contrary, so evident? Part of the answer may be because mimesis is a form of collective—or more precisely, to use the excellent term coined by Girard, a form of *interdividual* behavior. Imitation always involves more than one agent. As Jean-Pierre Dupuy has argued, reciprocal imitation should be considered both as the most simple and as the paradigmatic form of imitation.[6] That is to say, mimesis, unlike the explicit imitation of one person by another, is never something that one agent does to another, but something that people do to each other, something that always involves reciprocally more than one person. If Proteus *only* imitated Valentine, events would have evolved quite differently,[7] but it is precisely because Valentine seeks his friend's approval that he presents Silvia to him and praises her so much. Valentine can only "really" desire her if Proteus also does. Valentine imitates Proteus just as much as Proteus imitates Valentine. What results is the consequence of their joint action, of which they are mostly unaware. Each one ignores how rigidly his behavior is determined by that of the other, and although they are in opposition, they collaborate to produce each other's actions.

Mimesis is interdividual in the sense that it is a way through which agents determine each other's desires or intentions.[8] This reciprocal influence takes place at a pre-individual level. In this sense, mimesis takes place at a level at which agents are not subjectively aware of the influence they exert on each other, though afterwards they may become conscious of it through inference. It also takes place at a pre-intentional level. Even if it is part of Valentine's intention to get Proteus's approval for his choice of Silvia, even if he may be slightly aware that he is praising her excessively in an attempt to win over his friend, even if he may obscurely desire that his friend to some extent "falls in love with Sylvia," it seems unlikely that it is part of his intention for Proteus to become his rival. What is it that Valentine is at least vaguely conscious of here? According to Girard, Valentine is vaguely aware of his inability to desire autonomously, of his constant need to have his choices approved and confirmed by others. If there is no experience of mimesis, it seems that there is at least an experience of that. If Girard is right, this experience must be fairly universal, in the sense that everyone must have experienced it, more or less clearly, at least sometime. If asked, most would be ready to associate a certain feeling with this experience of needing the approval and confirmation of others—probably one that is not entirely comfortable. However, this frequent experience does not seem to have any specific expression that makes it public and transparent to others. What do people look like when they feel this way? There is no particular answer to that question. They look like everybody else. Neither someone who is presently experiencing this, nor those who are more often conscious of their dependence upon the choice of others seem to have any particular characteristics that clearly gives them away. This then, unlike an emotion, is a radically private experience. It is an experience that one can only make public and share with others as the result of a conscious decision. There is no built-in mechanism that renders this "mental state" visible to all.

STABILITY AND TRANSFORMATION

Why is this so? Perhaps is it better to begin by asking a different question: Why are emotions, unlike mimesis, characterized by determinate expression and subjective feelings? Emotions, or more precisely affective expression is, I submit, a means of intraspecific coordination through which agents determine each other's intention towards the other. Those events to which we give the name of emotions are salient moments within this ongoing process of coordination. Emotions, at least a few of them, have relatively standardized forms of expression because these salient moments play the role of conventions of coordination. They are also associated with strong subjective feelings that make agents selectively sensitive to the type of information that confirms the solution of coordination, which they constitute. For example, if I am afraid of you, I tend to interpret every aspect of your behavior as threatening. On the other hand, if I am in love with you, I become blind to all your failings. Emotions have evident expression because this expression is both the way in which coordination is reached and a sign that it has taken place. They come with strong subjective experiences because these tend to stabilize the solutions of coordination, which these salient moments constitute. On the other hand, mimesis does not have any form of expression that is associated with it, nor does it lead to any particularly strong subjective feeling. In other words, whether love, hate, envy, spite, dejection, admiration, or resentment, any and all emotions and passions can be related to mimesis; no single strong subjective feeling in particular is mimesis. In that sense, mimesis seems much closer to the process of intersubjective exchange that determines emotions, rather than to emotions as such, which constitute salient moments within that process—moments characterized by evident expression and strong subjective feelings. Like mimesis, that process is also pre-individual or interdividual, and largely unconscious.[9]

In fact, I suspect that both mimesis and affective coordination are part of a very general mechanism that puts agents in communication at a pre-personal level, what Vittorio Gallese has named the "shared manifold."[10] To state it in one very short formula, what I take to be the main (functional) difference (and similarity) between affective coordination and mimesis within this process is that if both are dimensions or aspects of a process through which the intentions of agents toward each other are determined, affective coordination is the means through which their intentions towards each other are arrested, while mimesis is the process through which they change. That, at least, is the hypothesis that I wish to explore in the remainder of this paper.

Valentine and Proteus are friends. This friendship is a guarantee of stability in their relation. The more they like each other, the more we expect their relationship to last. We assume that in difficult situations they will help each other, that they can safely trust each other with their most intimate secrets. Of course we also know that this is not entirely true, and if we do not, Girard and Shakespeare are there to remind us of how fragile "true" friendship can be. However, neither is it entirely false that those who are friends do help and trust each other. As I mentioned earlier, the expression of emotion should not always be taken at face value; nonetheless it is part of its essence that it has that face value. I suspect that this determination of future behavior through what emotions announce would always be true if emotions (or affective coordination) were all that there is. If such were the case, as it is in the world described in the pastoral genre according to Girard,[11] friends and lovers would be faithful to each other, and enemies would hate and be at odds permanently; the world would be a well ordered and structured place in the sense that standard expectations related to different emotions and situations would always be satisfied. Affective coordination is the means through which these standard expectations are constructed, and to some extent satisfied, by determining each agent's intention toward the other.

What Girard, in his analysis of *A Midsummer Night's Dream*, describes as the work of mimetic desire is the exact opposite of this process. First, Hermia tells her friend Helena her intention to elope with Lysander, with whom she is in love. Helena reveals this secret to Demetrius, who loves Hermia, but with whom she, Helena, is in love. All four soon leave for the forest, and almost immediately after, Lysander abandons in the woods the woman who entrusted her life and reputation to him, in order to pursue Helena with whom he has in the meantime fallen in love. Demetrius rapidly follows suit and also falls in love with Helena; then Hermia tries to seduce Demetrius, with whom Helena, her best friend, is in love, and so on, until

> as the end approaches, the metaphysical absolute shifts from character to character and the mimetic relation loses all stability. When the two boys abandon Hermia and turn to Helena, the entire configuration is reorganized on the basis of the same polarities, but with a new distribution of roles. A formerly despised member of the group has become its idol, and a former idol has lost all prestige.[12]

It is not only "mimetic relations" that lose all stability; differences between emotions and agents change so rapidly that they lose all consistency. What Girard describes is a mechanism through which the differences between agents that definite emotions establish, as well as the expectations they make possible, disappear.

EXPECTATIONS AND CHANGE

In *Émotions*, the paradigmatic situation I describe is that of face-to-face encounter.[13] As a mechanism of intraspecific coordination, emotions function first of all at the level of actual physical presence. It is through perceived changes in bodily posture and facial expression that agents act upon each other to determine each other's intention towards the other. At this level, the phenomenon is continuous and has an immediacy in which the salient events of coordination, which we call emotions, introduce discontinuity. Their salience in terms of evident expression and of strong subjective experience creates discontinuity in the flux of affective exchange. They constitute a point when the intentions of agents toward each other become determined. When that happens, the ongoing process of coordination is momentarily overruled and interrupted. There are good biological (evolutionary) reasons to believe that it would be advantageous if at any time, in spite of the anticipation of the other's behavior that emotions constitute, it would be possible to resume the ongoing process of coordination. That would allow agents to be able to react in real time to changes in their partner's attitude. Anecdotal evidence suggests that this is actually the case.

At the level at which this mechanism takes place, successful coordination does not involve preference for any specific solution of coordination. Whatever emotions arise—whether they be love, mutual hate, anger or submission, spite or resentment—equilibrium has been found in the sense that agents have formed an expectation concerning each other's intention towards one another. The ultimate reason why this mechanism shows no preference for one solution of coordination over the other is because this process constitutes the means through which expectations and preferences are primarily formed. Its goal is to allow agents to form expectations, but it does not have any preference for the content of the expectations that are created. I take it that this is also true of the agents before their expectations have been formed. In other words, even though I may have a general preference for getting along well with everyone, once I have met him or her, whether he or she turns out to be a jerk, an idiot, a pompous windbag, or someone I really, really like is not something about which I had any preference, at least in the sense that no expectation has been frustrated.

However, once expectations have been made, agents do have preferences concerning the future behavior of other agents. Their first preference is for their expectation to be satisfied in the future. We expect our friends to remain our friends, and our enemies to remain our enemies, and at a higher level of "meta-preferences" we prefer things to remain that way. That is to say, we prefer for others to adopt the behavior that their affective expression makes us anticipate they will have in the future. We tend, however, to be much more liberal concerning ourselves. We often do not see the fact that we may change as particularly shocking or disturbing.[14]

Mimesis, as Girard describes it, intervenes once these expectations have been made, once this first level of coordination has been reached. Face-to-face relationship is of course fundamental, but stability in relation requires the ability, so to speak, to "jump over" periods of absence. Of course this is precisely what emotions provide: expectations. Valentine desires and expects Proteus's approval of his love for Sylvia. He also fears a little that he may not get it, and, if he were just slightly more complex a character, he might also fear a little that he could be too successful in getting his friend to see his love favorably. The description of mimesis assumes that the language of emotion is already given, that agents have already formed expectations concerning each other's behavior toward the other, and it describes the process through which these expectations *will not be fulfilled*. That is to say, the process through which agents' intentions of action will come to change. Girard also claims, as we have seen at the beginning of this paper, that it is through imitation (or mimesis?) that these expectations have in the past been fulfilled. Following Shakespeare, he argues that Proteus's and Valentine's expectations of continued friendship have many times been satisfied in the past and reinforced through the very process of being satisfied. He adds, but does not really analyze or demonstrate, that the continued satisfaction of those expectations "really depends on a mutual *imitation* so spontaneous and constant that it remains unconscious."[15] Does it depend on imitation? Is this "imitation" the same thing as mimesis?

In other words, is the mechanism through which our expectations of each other's behavior are satisfied the same as that through which they are thwarted? The answer to that question is perhaps "yes," but that question is different from that of knowing how and why agents come to create such expectations. That is the question which the theory of emotions pretends to answer. Where things become difficult to disentangle is that Girard claims that agents are not individually in charge of the process through which their desires or intentions are determined. Mimesis is not something that belongs to one agent alone; it only arises in interrelations. This process is spontaneous, constant, and unconscious. I believe the same may be said about the process of affective coordination: that it is spontaneous, constant, and unconscious, and that it takes place at a subpersonal level. The main difference is that mimesis is a process that results in "convergence"—or if that term is already too charged normatively, mimesis is geared towards producing symmetry and similarity in behavior, and it does. Notwithstanding the fact that we do not generally perceive this symmetry, the symmetry is there.

In order to perceive this symmetry, we need to step out from where the agents involved in the interrelation are standing, and look at their behavior—from the position alluded to earlier, from the third-person point of view. It is not as if symmetry and similarity never appear to the agents, but the fact is, it is also there when they don't see it. Those who hate each other and fight are doing the "same thing" just as much as those who love and agree. Affective coordination, on the other hand, normally results in a finite set of (conventional) determined (differentiated) attitudes. Mimesis acts upon this set of differences. It progressively transforms love into hate, and in the ultimate stages makes love and hate indistinct.

In its social dimension, mimetic theory, through the scapegoat mechanism, is almost perfectly self-sufficient, as mimetic relations both destroy and create social differences. However when we come to individual psychology, this does not seem to be the case. Girard, in the context of psychology, seems much more interested in tracking the process through which differences are lost than that through which they are created. A quick look at the table of contents of *Des choses cachées depuis la fondation du monde* (*Things Hidden since the Foundation of the World*) tends to confirm this impression. The first part of the book, dedicated to anthropology, has sections on the genesis of myths, rituals, and social organization, that is to say on the creation of social differences. Part 3, which is dedicated to interdividual psychology, essentially reads as a treatise of psychopathology, an essay on the disappearance of psychological differences.[16] Unlike what happens in the case of societies, which live through a sacrificial crisis, neither in Girard's writing nor in our ordinary experience does it seem that agents who travel to the end of psychic dissociation through exceedingly intense mimetic relations will be "born again" on the other side and endowed with a new set of stable differences. There is, however, apparently one exception to this rule: some novelists—those who reveal, rather than simply reflect the importance of mimetic desire in human relations. Girard often compares the experience from which the good, *romanesque*, as opposed to the bad, romantic novel is born, to a religious conversion.[17] He does not say very much about it, but he suggests that this ability to *see* mimesis is given to those who have traveled quite some way down the path of psychological destruction to which it leads. This may be the case; however it is very clear that, according to him, there are also many who go very far in that direction and in fact never learn.

It seems, therefore, that in the psychological case, differences need to be given from the start. They are given, I believe at least in part, through the mechanism of affective coordination. It is clear, furthermore, that this mechanism is culturally constrained, and through this we return to the differences that are generated through mimetic interaction.

NOTES

1. René Girard, *A Theater of Envy: William Shakespeare* (New York: Oxford University Press, 1991), 9.

2. That is why I personally find the whole discussion concerning "good" and "bad" mimesis rather confused and misleading.

3. This does not mean that the relation between emotion and behavior is necessarily more direct than that between mimesis and behavior. In fact, once you have perceived the role of imitation in human affairs, you rapidly see that its relation with agents' choices and decisions is quite direct. The point, rather, is that we usually are not aware of the mimetic dimension of our action.

4. In Japanese spoken language, for example, the word "narau" means both to learn and to imitate. Written language, however, gives different *kanji* radicals to the two verbs.

5. Richard W. Byrne, "Animal Imitation," *Current Biology* 19 (2009): 111–14; Richard W. Byrne, "Parsing Behavior: A Mundane Origin for an Extraordinary Ability," in *The Roots of*

Human Sociality, ed. N. Einfield and S. Levinson (Oxford: Berg, 2006), 478–505; Andrew Meltzoff and Wolfgang Prinz, *The Imitative Mind: Development, Evolution, and Brain Bases.* (Cambridge: Cambridge University Press, 2002).

6. Jean-Pierre Dupuy, "Naturalizing Mimetic Theory," this volume.

7. For example, in *As You Like It* where Celia declines to take the bait offered her by Rosalind, who declares to her friend her love for Orlando and seeks her approval. Celia approves but does not fall in love with Orlando.

8. See Jean-Michel Oughourlian, "From Universal Mimesis to the Self Formed by Desire" and Jean-Pierre Dupuy, "Naturalizing Mimetic Theory," both in this volume.

9. Don Ross and Paul Dumouchel, "Emotions as Strategic Signals," *Rationality and Society* 16 (2004): 251–86; Paul Dumouchel, "Social Emotions," in *Animating Expressive Characters for Social Interaction*, ed. Lola Canamero and Ruth Aylett (Philadelphia: John Benjamin Publishing Company, 2008), 1–19.

10. Vittorio Gallese, "'Being like me': Self-Other Identity, Mirror Neurons, and Empathy," in *Perspectives on Imitation*, ed. Susan Hurley and Nick Chater (Cambridge: Cambridge University Press, 2005), 101–18; Vittorio Gallese, "The 'Shared Manifold' Hypothesis: From Mirror Neurons to Empathy," *Journal of Consciousness Studies* 8 (2001): 33–55. See also Vittorio Gallese, "The Two Sides of Mimesis: Mimetic Theory, Embodied Simulation, and Social Identification," this volume.

11. René Girard, *A Theater of Envy*, ch. 10.

12. Ibid., 51.

13. Paul Dumouchel, *Émotions: Essai sur le corps et le social* (Paris: Les Empêcheurs de penser en rond, 1999).

14. This, however, may be highly culturally relative.

15. René Girard, *A Theater of Envy*, 9.

16. René Girard, *Des choses cachées depuis la fondation du monde* (Paris: Grasset, 1978).

17. René Girard, *Deceit, Desire, and the Novel: Self and Other in Literary Structure*, trans. Yvonne Freccero (Baltimore: Johns Hopkins University Press, 1965).

CHAPTER 5

The Two Sides of Mimesis: Mimetic Theory, Embodied Simulation, and Social Identification

Vittorio Gallese

INTRODUCTION

René Girard (b. 1923), French literary critic and anthropologist, has provided us with an incredibly rich and thought-provoking theory of human culture: Mimetic Theory. What is most fascinating in Girard's Mimetic Theory is its broad and bold scope. According to Girard, human culture sits on the shoulders of religion, which in turn stems from the ritualization of social violence through the mechanism of scapegoating. As Girard wrote in *Violence and the Sacred*, "My theory is the first to offer an explanation of the primordial role that religion plays in primitive societies, as well as of man's ignorance of this role."[1]

Crucial in Girard's theory is the notion of *mimetic desire* as the main source of aggressiveness and violence characterizing our species. In "Mimesis and Violence," Girard writes: "It seems to me that a theory of conflict based primarily on appropriative mimicry does not have the drawbacks of one based on scarcity [of resources] or on aggressiveness; if it is correctly conceived and formulated it throws a great deal of light on much human culture, beginning with religious institutions."[2]

What is the *appropriative mimicry* Girard refers to, and where does it come from? It is the compulsive tendency of mankind to imitate others' desires, so that what is really desired and sought out is whatever is desired and sought out by others. The intrinsic value of the objects of our desire is not as relevant as the fact that the very same objects are the targets of the desires of others. To further spell it out in Girard's own words, "Violence is the process itself when two or more partners try to prevent one another from appropriating the object they all desire through physical or other means."[3]

The novelty of Girard's approach with respect to more traditional accounts of human violence is worth noting. Girard is very clear in drawing a distinction between desire and appetite.[4] While the latter is the outcome of instinctual drives, the former, typically, not only requires an object, but also another individual, the model or mediator. Mimetic Theory therefore proposes itself as a key option for solving the problem of *social cognition*.[5]

An objection could in principle be raised against the apparently negative and one-sided view of mankind, in general, and of mimesis, in particular, stemming from Girard's theory. Aren't human beings, after all, equally describable as empathic creatures, capable of fellow feelings, love, and altruism? Furthermore, one could argue that mimesis not only generates violence, but also art, culture, and creativity.

However, although there is no doubt that Girard's emphasis is mostly on human violence, the above-mentioned arguments would unfairly misrepresent Girard's thought. Girard acknowledged in his work, though perhaps with less emphasis, that mimetic desire is also good in itself because it is the basis of love, viewed as the imitation of a positive model.[6] Even more importantly, in my opinion, Girard stresses that "mimetic desire, even when bad, is intrinsically good, in the sense that far from being merely imitative in a small sense, it's the opening out of oneself."[7]

It is from this point—the notion of desire as openness to others—that I would like to start discussing some of the implications of the notion of mimesis against the background of Girard's Mimetic Theory. I will do so in order to show how empirical research in neuroscience and developmental psychology can shed new light on intersubjectivity, a crucial aspect of the human condition. Girard's Mimetic Theory constitutes an ideal starting framework to foster a multidisciplinary approach to this crucial topic.

Capitalizing upon aspects of the work of Alexandre Kojève, Martin Heidegger, and Helmuth Plessner, I will first illustrate how we can envisage a different, complementary, not mutually exclusive account of mimesis as one of the driving forces leading to social identification, and hence to human sociality and intersubjectivity. I will subsequently present a concise survey of empirical research in neuroscience and developmental psychology showing that this account of mimesis finds solid supporting evidence. A neuroscientifically based model of intersubjectivity, the shared manifold of intersubjectivity and its underpinning functional mechanism *embodied simulation*, will be discussed in relation to social identification and mutual recognition.[8]

I will argue that social cognition must not be uniquely conceived of as metacognition relying on the use of the propositional attitudes of folk psychology. I will argue that folk psychology is not the sole account of interpersonal understanding. Perhaps not even the most relevant. Before and below metarepresentational mind reading is *intercorporeity*—the mutual resonance of intentionally meaningful sensory-motor behaviors—as the main source of knowledge we directly gather about others.[9]

Intercorporeity describes a crucial aspect of intersubjectivity not because the latter is to be viewed as phylogenetically and ontogenetically grounded on a merely perceived similarity between our body and the body of others. Intercorporeity describes a crucial aspect of intersubjectivity because humans share the same intentional objects, and their *situated* motor systems are similarly wired to accomplish similar basic goals. Before and below our theoretical take on the world is the pragmatic character of our openness to the world.[10] I will argue that human beings are primarily wired to identify with each other, and that such a process of identification can be neurally grounded since the discovery of mirror neurons and other mirroring neural mechanisms.

Finally, capitalizing upon the empirical evidence here reviewed, I will come back to Girard's notion of mimetic desire, and propose that such a notion can fully exploit its heuristic value only by taking into account the fact that the primary object of desire is the "other." It will be concluded that a thorough and biologically plausible account of human social cognition requires the integration of both sides of mimesis.

MIMETIC THEORY, DESIRE, THE BODY, AND THE OTHER

Alexandre Kojève (1902–1968) can be viewed as an anticipator of the notion of mimetic desire. In his *Introduction to the Reading of Hegel*, he introduces the notion of the "desiring I" as a void to be filled by the positive content stemming from the action that by negating and destroying the desired "nonself," assimilates it.[11] Eating to satisfy hunger is one example of this type of self/non-self interaction. This condition, though, is not uniquely human, but shared with the animal world.

Human desire, though, can exist as such only within a plurality of other desires, that is, within a society of desiring human beings. In fact, when desire is targeting real material objects, it is *human* only to the extent that it is *mediated* by the desire of others targeting the same object. Kojève writes: "It is human to desire what others desire because they desire it."[12] In Kojève's view, human history is the history of desired desires.

Besides noticing the proximity of this view with Girard's notion of mimetic desire, I think it is important to stress that according to Kojève, the desire that defines the human condition is the desire directed towards another desire. It is only through this type of interaction that self-consciousness can be achieved. In fact, the object of this type of desire is non-natural, because this object, another desire, or better, the desire of someone else, is different from any material thing. A desire before its fulfillment is an oxymoron, nothing but the presence of an absence of reality.

The desire of being the target of others' desire becomes one of the distinctive features of the extreme alterity of humanity from nature and vitality, one of the main themes of Kojève's phenomenological anthropology. To desire another's desire, to be the target of others' desire, means to gain *social recognition*.

It does not interest us here where this notion of desire led Kojève. What is important for our discussion of mimesis and mimetic desire is the fact that the plurality of the mimetic desires of humans is strictly intertwined with the issue of social identification and recognition.[13]

This issue surfaces at different times in Western philosophical thought at the beginning of the twentieth century, and notably, among others, in the work of Martin Heidegger[14] and Helmuth Plessner.[15] Both philosophers, although starting from different premises, underline the pragmatic nature of the human condition and criticize the subject-object dichotomy of traditional ontology.

For Heidegger, the facticity of human existence shapes reality as a field of pragmatic meanings.[16] Our relation to the world of things and other individuals is pre-theoretical, as it stems from an original openness to the world, synthesized within the notion of *cure* (*Sorge*), literally, taking care of.[17] The peculiar condition of human nature, according to Heidegger, can be characterized as "being-in-the-world," that is, being inherently related to the world through our origin in being emotionally situated, expressed by his notion of *Befindlichkeit*.[18] Our comprehension of reality always bears an emotionally situated character, and it is a precondition for any detached, abstract theoretical analysis of the world, which thus acquires a secondary and derived character.

The intrinsic historical dimension of mankind with its constitutive projection into an open and problematic future forces man to reduce the number of possible decisions by relying on what others decide for us within the frame of consolidated social habits.[19]

Particularly pertinent is also the position of Helmuth Plessner (1989–1985) on the role of the body in intersubjectivity. According to Plessner, human beings, unlike other animals, are *eccentrically positioned* because they not only *are* a body but they also *possess* it. Such an eccentric position qualifies the human relation with itself as well as that with others, and in doing so enables it to go beyond the Cartesian dualism of body and soul.[20] Man is not only at the center of things, seen from a *hic et nunc* egocentric perspective, as animals are. Man entertains at the same time a reciprocal bond with himself and with others.

Plessner's take on intersubjectivity stems from this peculiar view on human eccentric positionality. In Plessner's view, the constitution of the *I* within a given body is anticipated and predetermined by the dimension of the *Thou* and of the *We*. Man's knowledge that he is neither just a thing among other things, nor alone in the world, but connected to a social community of other human beings, is not the outcome of an act of projection outside his life form, but springs from the intrinsic reality of human existence. In other words, intersubjectivity is a constitutive and fundamental element of the human condition.

This brief and sketchy overview suggests a tight relationship between human sociality and the natural and intrinsic pragmatic relatedness we entertain with the world, on the one hand, and our constitutive—ontological—relatedness to others. In the next sections we will explore what contemporary neuroscience and developmental psychology have to say on these matters.

AN EPISTEMOLOGICAL OVERTURE: OF NEURONS AND PERSONS

Before addressing these topics, we should first ask—and possibly answer—a preliminary epistemological question: How can neuroscience possibly shed light on personal-level issues given its peculiar epistemological approach consisting of a subpersonal level of description?

The standard approach of contemporary neuroscience to the problem of social cognition is indeed exposed to what Bennett and Hacker defined as the "mereological fallacy," that is, to attribute to one part of a living organism—e.g. the functional properties of the nervous system—characteristics that are proper to the entire organism as a whole.[21] Mentalization and intersubjectivity are competences uniquely describable at the personal level, and therefore not entirely reducible to the subpersonal activation of neural networks in the brain, hypothetically specialized in mind reading, as too many neuroscientists nowadays think. Neurons are not epistemic agents.[22] The only things neurons "know" about the world are the ions constantly flowing through their membranes. In contrast, mentalization and intersubjectivity are personal-level properties of individuals. We could tentatively define individuals as interconnected brain-body systems interacting in situated ways with a specific environment—our *Umwelt*—inhabited by other brain-body systems.

To make things worse, such an epistemological attitude is often combined with a blind reliance on brain-imaging techniques like fMRI as the sole method of investigation. This appears to be a highly risky enterprise: fMRI, if not supported by a detailed phenomenological analysis of the investigated perceptive, motor, and cognitive processes, and if not interpreted on the basis of the direct study of the activity of single neurons in the animal model, loses much of its heuristic power. The heuristic power of this approach is further reduced by the instrumental use of empirical data to validate a preconceived model of the mind, considered true a priori. Such a model, most of the time, is the one proposed by classic cognitive science, according to which social cognition only consists of metacognition and the use of the propositional attitudes of folk psychology. It is highly questionable that this model fully captures the real essence and functional architecture of the human mind.

That said, we must stress that the solution to the mereological fallacy cannot consist in an undifferentiated form of holism. Rather, by means of a careful empirical analysis of the subpersonal mechanisms investigated by neuroscience, we can discover the multilayered character of the experience we make of the world. Even if such layers as clarified by neuroscience do not fully exhaust this experience, they enable a description of its genesis and structure. These data, in turn, can fuel and promote a renewed philosophical analysis. This is one of the main reasons why I think that a dialogue between cognitive neuroscience and philosophy is not only desirable but also necessary.

WHERE DOES MIMESIS COME FROM? NEUROSCIENTIFIC EVIDENCE

One of the cornerstones of Girard' s Mimetic Theory is the triangular relation between two individuals and the object of their acquisitive desire. How does each protagonist of this triangular relation understand that the "other" also wants the same object? What are the mechanisms enabling appropriative mimesis? How can each of the human vertexes of the Girardian mimetic triangle realize that the object, the third vertex, is the target of the other's purposeful action? The same question can be reformulated in the following way: How do people understand the goals and intentions of the actions of others?

We are beginning to understand what the neural mechanisms are that enable this peculiar quality of human nature. In the early 1990s a new class of premotor neurons was discovered in the premotor cortex of the macaque monkey brain. These neurons were labeled "mirror neurons."[23] Mirror neurons fire both when the monkey performs goal-directed motor acts, like grasping objects with the hand and/or the mouth, and when it observes similar acts performed by others. It was proposed that through the activation of these neurons, a direct form of action understanding is accomplished. The observed behavior is pre-reflexively understood because it is constituted as a goal-directed motor act in virtue of the activation in the observer's brain of the neurons presiding over the motor accomplishment of similar goals.

This account of action understanding is further corroborated by a recent discovery: the motor system of primates is functionally organized in terms of goal-directed motor acts, and not in terms of movements. In a recent study by Umiltà et al., hand-related neurons were recorded from premotor area F5 and the primary motor cortex (area F1) in monkeys trained to grasp objects using two different tools: "normal pliers" and "reverse pliers."[24] These tools require opposite movements to grasp an object: with normal pliers, the hand has to be first opened and then closed, as when grasping is executed with the bare hand, whereas with reverse pliers, the hand has to be first closed and then opened. The use of the two tools enabled the dissociation of neural activity related to hand movement from that related to the goal of the motor act.

All tested neurons in area F5, and half of neurons recorded from the primary motor cortex discharged in relation to the accomplishment of the goal of grasping—when the tool closed on the object—regardless of whether in this phase the hand opened or closed, that is, regardless of the movements employed to accomplish the goal. Since mirror neurons share this property with all other F5 grasping-related neurons, they can directly map the behavior of others in terms of goal-related motor acts. Thus it appears that goal coding is first and foremost not an abstract, mentalist, and experience-independent property, but a distinctive functional feature upon which the cortical motor system of nonhuman primates is organized. Goal-directed motor acts are the nuclear building blocks around which action is produced, perceived, and understood.[25]

After the discovery of mirror neurons in the macaque monkey brain, several studies using different experimental methodologies and techniques have also demonstrated that in the human brain, the neural circuits underpinning action execution directly map its perception when executed by others. These parieto-premotor cortical networks are defined as the Mirror Neuron System (MNS).[26]

During action observation, there is a strong activation of premotor and posterior parietal areas, the likely human homologue of the monkey areas in which mirror neurons were discovered and described. The MNS for actions in humans is somatotopically organized, with distinct cortical regions within the premotor and posterior parietal cortices being activated by the observation/execution of mouth-, hand-, and foot-related acts.

Even more relevant to our discussion of mimesis is the discovery that the MNS in humans is directly involved in imitation of simple movements,[27] and in the imitation learning of complex skills.[28] Many interesting phenomena described by social psychologists, like the "chameleon effect"—the unconscious mimicry by the observer of postures, expressions, and behaviors of her/his social partners[29]—with the MNS can find a neurophysiological explanation. It is worth noting that these instantiations of unconscious mimesis all share a pro-social character, because their occurrence tends to increase during social interactions with affiliative purposes.

The perception of communicative actions,[30] and the detection of basic action intentions,[31] activate the MNS as well. Furthermore, the premotor cortex containing the MNS is involved in processing action-related words and sentences,[32] suggesting that the MNS, together with other parts of the sensory-motor system, indeed plays a relevant role in language semantics.[33]

In sum, the MNS is a good candidate for the subpersonal instantiation of what enables appropriative mimesis. This, however, is only one side of mimesis. As I have argued before, another important aspect of mimesis is that of enabling social identification and mutual recognition. To that purpose, a mechanism capable of mapping actions and intentions, though highly relevant, would not suffice. There are other dimensions to be mapped before the status of "another self" can be attributed to others. These dimensions encompass emotions and sensations. The news is that emotions and sensations also appear to be mapped according to the same resonance mechanisms already addressed for the domain of action.

In fact, other mirroring mechanisms seem to be involved with our capacity to share emotions and sensations with others.[34] When perceiving others expressing emotions by means of their facial mimicry, the observers' facial muscles activate in a congruent manner, with intensity proportional to their empathic nature. Both observation and imitation of the facial expression of emotions activate the same restricted group of brain structures, including the ventral premotor cortex, the insula, and the amygdala.

Finally, in an fMRI study, the issue of how the I-Thou experience of a particular emotion is mapped in the human brain was specifically addressed. To that purpose, the brain activity of healthy participants was investigated during the phenomenal experience of disgust by having them inhale disgusting odorants. The same participants in the same experiment were also brain-scanned during the observation of disgust as displayed in video clips of other individuals dynamically expressing disgust with their facial expression. The results showed that witnessing the facial expression of disgust of others activates the left anterior insula at the same location activated by the first-person subjective experience of disgust.[35]

Both the production and the perception of emotion-related facial expressions or body postures impinge upon common neural structures related to viscero-motor, somato-motor, and affective aspects of the emotion experience. It appears, therefore, that there is a "we-centric" dimension in the experience of a given emotional/affective state, and that it is underpinned by the activity of a common neural substrate.

When we witness a given facial expression and comprehend that expression as characterized by a particular emotional state, we do not accomplish this type of comprehension through explicit inference from analogy. The other's emotion is first constituted and directly understood by means of embodied simulation producing an "as if" experience engendered by a shared body state. It is the body state shared by the observer and the observed that enables direct understanding.

This view appears to be congruent with the perspective proposed by Mead, and more recently by Paul Dumouchel, which considers emotions primarily from the output or expressive side.[36] As argued by Dumouchel, the universality of emotions reside in the universality of the social bonds they help constitute. According to Dumouchel, being in a given emotional state (say, being angry at someone) is not an intrinsic psychological property of a subject, but the relational property of an individual within a given social context. This means that the expression of emotions constitutes a system of social communication with the main purpose of facilitating social coordination.

Facial and bodily movements reveal preferences about available behavioral options.

Similar direct mapping mechanisms have been described for the perception of pain and touch.[37] Altogether, these results suggest that the same neural circuits underpinning our own actions, intentions, emotions, and sensations also underpin our capacity to recognize and identify with the actions, intentions, emotions, and sensations of others. Recent studies suggest that these mechanisms could be deficient in individuals affected by autistic spectrum disorders.[38]

The specific social cognitive flexibility of our species, as reflected in our propensity for pedagogy, and in the sophisticated quality of our social understanding, likely exceeds the functional properties of the MNS. However, I posit that a proper development of the MNS is a necessary prerequisite for scaffolding the development of the proper human social cognitive skills leading to mutual recognition and social identification.

WHERE DOES MIMESIS COME FROM? EVIDENCE FROM DEVELOPMENTAL PSYCHOLOGY

At the onset of life, interpersonal relations are readily established within a primitive shared "we-centric space."[39] Neonates share this space with their caregivers. The physical space occupied by the body of the caregiver—the mother, in the first place—is "hooked up" to the body of the infant to compose a shared space. This we-centric space becomes richer and multifaceted, due to the wider range and meaning of interpersonal relations in the course of development.

At birth, humans already appear to be engaged in interpersonal mimetic relations. The seminal study of Meltzoff and Moore and the subsequent research field it opened showed that newborns are capable of reproducing mouth and face movements displayed by the adult they are facing.[40] The particular part of their body responded, though not in a reflexive way, to movements displayed by the equivalent body part of someone else. As Meltzoff recently wrote, "The bedrock on which commonsense psychology is constructed is the apprehension that others are similar to the self. Infants are launched on their career of interpersonal relations with the basic perception: 'Here is something like me.'"[41] These results suggest that neonates are innately prepared to link to their caregivers through imitation, clarifying yet another of the various capacities that locate human infants in the social world from the very beginning of life.

Very early on, infants show unequivocal signs of social interaction sequences. They actively solicit their caregivers' attention and engage themselves in body activity displaying "protoconversational" turn-taking structure—that is, characterized by a structure remarkably similar to adult conversations.[42]

Trevarthen recently defined these early mother-child interactions as "primary musicality," where "protoconversations and games with infants carry narratives in cycles of effort and excitement, with predictable harmonies and pauses, and the infant anticipates the steps and remembers the distinctive melodies and rhyming cadences well, becoming an increasingly skilled co-performer."[43]

Furthermore, as beautifully shown by Reddy, preverbal infants only a few months old, when engaged in social interactions, even show signs of so-called "self-conscious emotions" like embarrassment, pride, and coyness at a developmental age preceding the onset of self-reflective consciousness, definitely well before they are capable of self-recognition when looking at their reflection in a mirror. As Reddy writes, "Engaging with other minds is an emotional process from start to finish." Immediately after, she adds, "Rather than derive from conceptual development in the second year of human infancy, these [self-conscious] emotions exist in simple forms as ways of managing the exposure of self to other from early in the first year and are crucial for shaping the infant's emerging conception of self and other."[44]

These results suggest that prior to any triangular mimetic relationship, the main object of infants' mimesis is the affective behavior of the "other." In sum, as pointed out by Beebe et al., developmental psychology has shown that the mind begins as a shared mind. I posit that mirroring mechanisms and the functional mechanism they underpin—embodied simulation—are a crucial component of what makes our mind in the first place a *shared mind*.[45]

The shared we-centric space enabled by the activation of mirroring mechanisms is paralleled by the development of perspectival spaces defined by the capacity to distinguish self from other, as long as self-control develops. Infants progressively develop an agentive, subjective perspective on the world. However, such a process of personal identification anchored to an egocentric perspective contains and depends upon a contrastive element. "In the absence of reciprocity there is no alter Ego," writes Merleau-Ponty.[46]

It is not possible to conceive of oneself as a "self" without rooting this process of appraisal in an earlier stage in which sharing prevails. Also, in adulthood a shared manifold of intersubjectivity underpins, scaffolds, and enables our social transactions.

Why We Embody a We-Centric Space: The Shared Manifold of Intersubjectivity, Embodied Simulation, and Social Identification

How can we explain the ease with which we normally understand other people when interacting with them? The hypothesis put forward here is that the I-Thou relation provides the basic ground for our cognitive/affective development, hence for our truest and most intimate abilities as social individuals capable of mutual recognition and understanding.

The MNS, together with the discoveries of developmental psychology here concisely and partially summarized, provides a new empirically based image of intersubjectivity viewed first and foremost as *intercorporeity*. Intercorporeity, in turn, leads to social identification. Social behavior is not peculiar to humans. Nevertheless, central to all human social cultures of whatever complexity is the notion of *social identification* with the members of those cultures.[47] All levels of social interaction that characterize cognition in single individuals, in one way or another, intersect or overlap with the notion of mutual recognition and intelligibility, that is, with the notion of social identification.

Social identification can be articulated on different levels of complexity. As human beings, we implicitly "know" that we all share certain features: we have four limbs, walk in a certain way, act in particular ways, etc. People sharing the same culture will, for example, tattoo their bodies in a peculiar fashion, wear the same regimental necktie at fraternal meetings, or share political values, such as being against the death sentence, etc. Social identification is the membership fee all individuals pay in order to self-guarantee the sense of belonging to a larger community of other individuals.

Social identification is adaptive, because it grants the capacity to better predict the consequences of the ongoing and future behavior of other members of a given social group. The attribution of the status of "another self" to other individuals automatically contextualizes their behavior. This, in turn, reduces the variables to be computed, thus optimizing the employment of cognitive resources by reducing the "meaning space" to be mapped. By contextualizing content, identification reduces the information our brain is supposed to process. Beside—and likely before—the ascription of any mental content to others, we entertain a series of "implicit certainties" about the content-bearing individuals we meet.[48] These implicit certainties are constitutive of interpersonal relations, in that they deal with the sense of oneness, of identity with the other that enables the possibility to ascribe any content to the individuals with whom we are interacting.

Whenever we meet someone, we are implicitly aware of his/her similarity to us, because we literally embody it. Meltzoff and Brooks have suggested that the "like me" analogy between infant and caregiver is the starting point for the development of (social) cognition.[49] The "like-me" status, though, is neither the outcome of an inference by analogy nor the result of our conscious reflection on a perceived external similarity. Our social identification with others is a constitutive endowment of what it means to be human.[50] The I-Thou relation[51] is shaped by bidirectional interaction processes,[52] hence "self" and "other" are originally co-constituted. Infants use the observed behavior of their human partners as a mirror to gain more knowledge about themselves. But the same process also works the other way around: it enables infants to know about others.

It has been proposed that a *shared manifold* characterizes our interpersonal relations.[53] This term characterizes what happens when we witness the actions of others, or witness their overt behavior expressing the sensations and emotions they experience. Basically, it describes our capacity for direct and implicit access to others as subjects of experience, as we are. The shared manifold of intersubjectivity can be described at three different levels: the phenomenal level, the functional level, and the subpersonal level.[54]

The *phenomenal level* is responsible for the sense of social identification—of being part of a larger social community of persons like us—normally experienced during our encounters with others. This phenomenal state generates the peculiar sense of familiarity with other individuals, our intentional attunement to them, produced by the collapse of others' intentions into those of the observer.

The *functional level* can be characterized in terms of *embodied simulation* of the actions we see and/or of the emotions and sensations whose expression we observe in others.

The notion of simulation is employed in many different domains, often with different, not necessarily overlapping, meanings. Simulation is a functional process that possesses certain content, typically focusing on possible states of its target object. In philosophy of mind, the notion of simulation has been used by proponents of the simulation theory of mind reading to characterize the pretend state adopted by the attributer in order to understand another person's behavior.[55] Basically, according to this view, we use our mind to put ourselves in the mental shoes of others.

In opposition to standard accounts of simulation theory (ST), I qualify simulation as *embodied* in order to characterize it as a mandatory, pre-rational, nonintrospectionist process. The folk-psychological model of mind reading proposed by standard accounts of simulation theory,[56] which Gallese and Goldman utilized to frame the functional relevance of mirror neurons,[57] in my opinion *does not apply* to the nonrepresentational character of embodied simulation, as spelled out in Gallese's work,[58] and in the present article. The embodied simulation model is in fact an attempt to avoid folk psychology as the sole account of interpersonal understanding. Before and below mind reading is *intercorporeity* as the main source of knowledge we directly gather about others.[59]

A direct form of understanding others from within, as it were—intentional attunement—is achieved by the activation of neural systems underpinning what we and others do and feel. Parallel to the detached third-person sensory description of the observed social stimuli, internal nonlinguistic "representations" of the body states associated with actions, emotions, and sensations are evoked in the observer, *as if* he or she were performing a similar action or experiencing a similar emotion or sensation.

It must be stressed that the term "representation" is used here very differently from its standard meaning in classic cognitive science and analytic philosophy. It refers to a particular type of content generated by the relations that our situated and interacting brain-body system instantiate with the world of others. Such content is pre-linguistic and pre-theoretical, but nevertheless has attributes normally and uniquely attributed to conceptual content.

Finally, the subpersonal level of the shared manifold is instantiated as the activity of a series of mirroring neural circuits. The activity of these mirror neural circuits is, in turn, tightly coupled with multilevel changes within body states. Mirror neurons in monkeys and the MNS in humans instantiate a multimodal shared space for actions and intentions. As we have seen, other data show that analogous neural networks outside the motor system are at work to generate multimodal emotional and somatosensory "we-centric" shared spaces.[60]

To put it in simpler words, every time we relate to other people, we automatically inhabit a we-centric space, within which we exploit a series of implicit certainties about the other. This implicit and pre-theoretical, but at the same time contentful state enables us to directly understand what the other person is doing, why he or she is doing it, and how he or she feels about a specific situation.

This, of course, doesn't imply that we experience others the way we experience ourselves. The I-Thou identity relation constitutes only one side of the intersubjectivity coin. As posited by Edmund Husserl, and recently reemphasized by Dan Zahavi, it is the alterity of the other that guarantees the objectivity we normally attribute to reality.[61] Our lived experience (*Erlebnis*) of the "external" world is determined by the presence of other sentient agents.

It must be noted that the alterity character of others as we experience them also maps at the subpersonal neural level, because the cortical circuits at work when *we* act neither completely overlap nor show the same activation intensity as when *others* are the agents and we are the witnesses of their actions. The same logic also applies to emotions[62] and sensations.[63]

It must also be stressed that the functional mechanism of embodied simulation should not be conceived as a rigid, reflex-like input-output coupling. Several brain-imaging studies conducted on human beings have shown that the intensity of the MNS activation during action observation depends on the similarity between the observed actions and the participants' action repertoire. In particular, one fMRI study focused on the distinction between the relative contribution of visual and motor experience in processing an observed action.[64] The results revealed greater activation of the MNS when the observed actions were frequently performed with respect to those that were only perceptually familiar but never practiced.

Every instantiation of mirroring or interpersonal resonance—that is, embodied simulation—is always a process in which the behavior of others is metabolized by, and filtered through, the observer's idiosyncratic past experiences, capacities, and mental attitudes. Future research will have to focus on the role played by factors like specific personality traits, gender, professional expertise, etc., in modulating these neural mechanisms.

Why not employ the term *empathy* to characterize the basic level of relatedness and social identification described so far? It is certainly possible, provided that empathy is redefined along the extended lines I have suggested, but one must bear in mind that in doing so, there is a price to be paid. First, the use of the notion of empathy, by virtue of its different connotations employed in different contexts, systematically exposes one to misunderstandings.[65] Second, the notion of empathy nowadays almost exclusively refers to the emotional/affective aspects of interpersonal relations, thus without covering important aspects of interpersonal relations, such as actions and intentions, nowadays traditionally ascribed to the propensity of mankind for theorizing.

In contrast, by means of the shared manifold model, we can accommodate and account for most—if not all—different expressive behaviors enabling us to establish a meaningful link with others. This provides a unified account of important aspects and levels of description of intersubjectivity mainly viewed as intercorporeity.

Mimetic Desire Revisited

Let us finally go back to where we started from, that is, to Girard's powerful idea of mimetic desire. According to Girard, human beings imitate others' desires. This means that what really matters when we desire something is not the intrinsic value of the objects of our desire, but the fact that the very same objects are the targets of others' desire, where the others are to be conceived as *models* or *mediators*. We should not forget, however, that mimetic desire starts as desire to be the other. Indeed, Girard himself expresses the very same idea in his first major work, *Deceit, Desire, and the Novel*, where he writes, "The impulse toward the object is ultimately an impulse toward the mediator"; and later on, "Imitative desire is always a desire to be Another."[66] The desire of taking possession of any given object is always derivative of the desire to be as the model. It is only because of humanity's desire to be someone else that objects become potential targets of acquisition. In this respect, referring to Max Scheler, Girard writes, "Max Scheler himself is not far from the truth when he states in *Ressentiment* that the fact of choosing a model for oneself is the result of a certain tendency, common to all men, to compare oneself with others."[67]

Girard's perspective, by underlining the ambivalent character of mimesis with its potentiality to lead humans either to escalating violence or to symbolic-cultural transmission, provides a very stimulating, broad-ranging, and challenging contribution to our understanding of the evolution of human culture.[68]

The idea I am trying to put forward here is that at the origin of mimetic ambivalence is humanity's ontological openness to others. Our "ontological" desire to be as the "other," the model, stems from our ontological openness to the other, which, in turn, is determined by the fact that the other is already a constitutive part of the self. What follows is that we should abandon the Cartesian view of the primacy of the ego, and adopt a perspective according to which the other is co-originally given as the self. Both self and other appear to be intimately intertwined because of the intercorporeity linking them. Self-individuation is a process originating from the necessity of disentangling the Self from the we-centric dimension in which it is originally and constitutively embedded.[69] Our constitutive openness to others, of which mimesis is one of the main expressions, can be declined both in terms of social violence and social cooperation.

As recently emphasized by De Presteer following Merleau-Ponty, the body of intercorporeity is primarily perceived as a systematic means to go towards objects. This is the reason why, argues De Presteer, "the other is seen as a behavior and the 'I' is primarily a 'motor I.'"[70] A direct form of understanding others from within, as it were, is achieved by the activation of neural systems like the MNS underpinning what others and we do and feel. Our own acting body thus becomes the main source of information about others' behavior, and intersubjectivity is at its roots chiefly intercorporeity.

The main ambition of this paper was to show that today we can ground man's openness to others in neurobiological bases.

CONCLUSIONS

We have examined empirical results showing how interpersonal relations are made possible, in the first place, by resonance mechanisms that provide the common ground upon which the I-Thou relation can be established.

It could be tempting to use such evidence to assert the neurobiological basis of the supposed natural proclivity of humanity to sympathy, fellow feelings, goodwill, and altruism. I think we must resist such temptation and look at human nature as it really is, and not as we would like it to be. In this respect, Girard's Mimetic Theory is illuminating, because it shows that mimesis, when declined as mimetic desire, has the intrinsic potentiality of driving humans to aggression and violence.

Mimesis, as I have been trying to show throughout this paper, is neither intrinsically good nor bad. It is a basic functional mechanism at the core of our diversified social competencies and activities. Nevertheless, mimesis has two sides. Any serious neuroscientific attempt to shed light on the truest and deepest nature of the human condition cannot neglect either side. I posit that the empirical evidence briefly summarized here has the potential to stimulate future research that, driven by the currently available evidence, may shed further light on both sides of mimesis.

There is an original experience we make of other human beings, no matter what their ethnic, religious, socioeconomic, or cultural status is. Such original experience appears to be rooted in neural mechanisms connecting different brain-body systems as *human beings like us*.

In present historical times, characterized by the vehement resurgence of ethno-religious particularism, when *identity* and *difference* are key problems at a global scale as well as at the level of our local communities, to establish that the universal status of human beings is produced by social identification and mutual recognition, and that it is biologically grounded, shows the potential ethical relevance of neuroscientific research.

This is why I think it is important to address the most typical philosophical problem—who are we—from a multidisciplinary perspective that incorporates what disciplines like neuroscience and developmental psychology can teach us.

ACKNOWLEDGMENTS

This work was partly supported by MIUR (Ministero Italiano dell'Università e della Ricerca), by the EU grant NESTCOM, and the EU Marie Curie–Research Training Network 035975 "DISCOS—Disorders and coherence of the embodied self."

NOTES

1. René Girard, *Violence and the Sacred*, trans. Patrick Gregory (Baltimore: Johns Hopkins University Press, 1977), 310.

2. René Girard, "Mimesis and Violence: Perspectives in Cultural Criticism," *Berkshire Review* 14 (1979): 10.

3. Ibid., 9.

4. Girard, *Violence and the Sacred*.

5. As will become clearer as the paper develops, social cognition is indeed a problem, because divergent explanations are currently proposed both at the theoretical level by philosophy of mind and cognitive science, and at the level of the functional mechanisms supposedly underpinning it by cognitive neuroscience.

6. See René Girard, "Violence, Difference, Sacrifice: A Conversation with René Girard," *Religion and Literature* 25 (1993): 9–33.

7. Ibid., 24.

8. Vittorio Gallese, "Embodied Simulation: From Neurons to Phenomenal Experience," *Phenomenology and the Cognitive Sciences* 4 (2005): 23–48; Vittorio Gallese, "Before and Below Theory of Mind: Embodied Simulation and the Neural Correlates of Social Cognition," *Proceedings of the Royal Society B: Biological Sciences* 362 (2007): 659–69; Vittorio Gallese et al., "Motor Abstraction: A Neuroscientific Account of How Action Goals and Intentions are Mapped and Understood," *Psychological Research* 73, no. 4 (2009): 486–98.

9. Gallese, "Before and Below," 659–69; Gallese et al., "Motor Abstraction," 486–98.

10. Gallese, "Before and Below," 659–69; Gallese et al., "Motor Abstraction," 486–98; Vittorio Gallese, "Motor Cognition and its Role in the Phylogeny and Ontogeny of Intentional Understanding," *Developmental Psychology* 45 (2009): 103–13.

11. Alexandre Kojève, *Introduction à la lecture de Hegel: Leçons sur la Phénoménologie de l'Esprit professés de 1933 à 1939 à l'École des Hautes Études réunies et publiées par Raymond Queneau* (Paris: Éditions Gallimard, 1947).

12. "Il est humain de désirer ce que désirent les autres, parce qu'ils le désirent." Kojève, *Introduction*, 13.

13. For a recent discussion of social recognition, see Axel Honneth, *Verdinglichung* (Frankfurt am Main: Suhrkamp Verlag), 2005.

14. Martin Heidegger, *History of the Concept of Time* (Bloomington: Indiana University Press, 1925); Martin Heidegger, *Being and Time* (Albany: State University of New York Press, 1927); Martin Heidegger *The Fundamental Concepts of Metaphysics: World, Finitude, Solitude* (Bloomington: Indiana University Press, 1929); Martin Heidegger, *Grundbegriffe der Aristotelischen Philosophie*, vol. 18, part 2 of *Gesamtausgabe* (Frankfurt am Main: Klostermann, 2002).

15. Helmuth Plessner, *Die Stufen des Organischen und der Mensch: Einleitung in die philosophische Anthropologie*, vol. 4 of *Gesammelte Schriften* (Frankfurt am Main: Suhrkamp, 1928).

16. Heidegger, *Aristotelischen Philosophie*; Heidegger, *Concept of Time*; Heidegger, *Being and Time*; Heidegger, *Concepts of Metaphysics*.

17. Heidegger, *Being and Time*.

18. Ibid.

19. Ibid.

20. Plessner, *Die Stufen*.

21. M. Bennett and P. Hacker, *Philosophical Foundations of Neuroscience* (London: Blackwell Publishing, 2003).

22. Gallese, "Before and Below," 659–69.

23. Vittorio Gallese et al., "Action Recognition in the Premotor Cortex," *Brain* 119 (1996): 593–609; G. Rizzolatti et al., "Premotor Cortex and the Recognition of Motor Actions," *Cognitive Brain Research* 3 (1996): 131–41.

24. M. Umiltà et al., "How Pliers Become Fingers in the Monkey Motor System," *Proceedings of the National Academy of Sciences* 105 (2008): 2209–13.

25. Vittorio Gallese et al., "Motor Cognition and Its Role in the Phylogeny and Ontogeny of Intentional Understanding," *Developmental Psychology* 45 (2009): 103–13.

26. For a review, see G. Rizzolatti et al., "Neurophysiological Mechanisms Underlying the Understanding and Imitation of Action," *Nature Neuroscience Reviews* 2 (2001): 661–70; Vittorio Gallese, "The Manifold Nature of Interpersonal Relations: The Quest for a Common Mechanism," *Philosophical Transactions of the Royal Society of London-B* 358 (2003): 517–28; Vittorio Gallese et al., "A Unifying View of the Basis of Social Cognition," *Trends in Cognitive Sciences* 8 (2004): 396–403; Gallese, "Embodied Simulation," 2005; Vittorio Gallese, "Intentional Attunement: A Neurophysiological Perspective on Social Cognition and Its Disruption in Autism," *Exp. Brain Res. Cog. Brain Res.* 1079 (2006): 15–24; Gallese, "Before and Below," 659–69; Gallese et al., "Motor Abstraction," 486–98; Giacomo Rizzolatti and Laila Craighero, "The Mirror-Neuron System," *Annual Review of Neuroscience* 27 (2004): 169–92.

27. Marco Iacoboni et al., "Cortical Mechanisms of Human Imitation," *Science* 286, no. 5449 (1999): 2526–28; M. Iacoboni et al., "Reafferent Copies of Imitated Actions in the Right Superior Temporal Cortex," *Proceedings of the National Academy of Sciences, USA* 98 (2001): 13995–99; N. Nishitani and R. Hari, "Temporal Dynamics of Cortical Representation for Action," *Proceedings of the National Academy of Sciences, USA* 97 (2000): 913–18; N. Nishitani and R. Hari, "Viewing Lip Forms: Cortical Dynamics," *Neuron* 36 (2002): 1211–20.

28. G. Buccino et al., "Neural Circuits Involved in the Recognition of Actions Performed by Nonconspecifics: An fMRI Study," *Journal of Cognitive Neuroscience* 16 (2004): 114–26; S. Vogt et al., "Prefrontal Involvement in Imitation Learning of Hand Actions: Effects of Practice and Expertise," *Neuroimage* 37 (2007): 1371–83. In Theodore Lipps we find a first suggested relation between mimesis—conceived as "inner imitation" (*Innere Nachahmung,* in Lipps's words)—and the capacity of understanding others by ascribing to them feelings, emotions, and thoughts. Theodore Lipps, "Einfühlung, innere nachahmung und organen-empfindung," *Archiv. F. die Ges. Psy.* 1 (1903): part 2.

29. T. Chartrand and J. Bargh, "The Chameleon Effect: The Perception-Behavior Link and Social Interaction," *Journal of Personality and Social Psychology* 76 (1999): 893–910.

30. G. Buccino et al., "Neural Circuits Underlying Imitation Learning of Hand Actions: An Event-Related fMRI Study," *Neuron* 42 (2004): 323–34.

31. M. Iacoboni et al., "Grasping the Intentions of Others with One's Own Mirror Neuron System," *PLOS Biology* 3 (2005): 529–35; G. Buccino et al., "The Neural Basis for Non-Intended Actions," *Neuroimage* 36, no. 2 (2007): 119–27.

32. For a review, see Gallese, "Before and Below," 659–69; Vittorio Gallese, "Mirror Neurons and the Social Nature of Language: The Neural Exploitation Hypothesis," *Social Neuroscience* 3 (2008): 317–33.

33. See F. Pulvermüller, *The Neuroscience of Language* (Cambridge: Cambridge University Press, 2002); Vittorio Gallese and G. Lakoff, "The Brain's Concepts: The Role of the Sensory-Motor System in Reason and Language," *Cognitive Neuropsychology* 22 (2005): 455–79. See also note 32 above.

34. Vittorio Gallese, "The "Shared Manifold" Hypothesis: From Mirror Neurons to Empathy," *Journal of Consciousness Studies* 8, no. 5–7 (2001): 33–50; Gallese, "Manifold Nature," 517–28; Gallese, "Intentional Attunement," 15–24; F. de Vignemont and T. Singer, "The Empathic Brain: How, When, and Why?" *Trends in the Cognitive Sciences* 10 (2006): 435–41; J. A. Sommerville and J. Decety, "Weaving the Fabric of Social Interaction: Articulating Developmental Psychology and Cognitive Neuroscience in the Domain of Motor Cognition," *Psychonomic Bulletin & Review* 13 (2006): 179–200.

35. B. Wicker et al., "Both of Us Disgusted in My Insula: The Common Neural Basis of Seeing and Feeling Disgust," *Neuron* 40 (2003): 655–64.

36. G. H. Mead, *Mind, Self, and Society* (Chicago: Chicago University Press, 1934); Paul Dumouchel, *Émotions: Essai sur le corps et le social* (Paris: Le Plessis-Robinson, Institut Synthélabo, 1995); see also Paul Dumouchel, "Emotions and Mimesis," in this volume.

37. See Gallese, "Intentional Attunement," 15–24.

38. For review, see note 28 above; L. M. Oberman and V. S. Ramachandran, "The Simulating Social Mind: Mirror Neuron System and Simulation in the Social and Communicative Deficits of Autism Spectrum Disorder," *Psychological Bulletin* 133 (2007): 310–27.

39. Gallese, "Shared Manifold Hypothesis," 33–50; Gallese, "Manifold Nature," 517–28; Vittorio Gallese, "'Being Like Me': Self-Other Identity, Mirror Neurons and Empathy," in *Perspectives on Imitation: From Cognitive Neuroscience to Social Science,* ed. S. Hurley and N. Chater (Cambridge, MA: MIT Press, 2005), 1:101–18; Vittorio Gallese, "Embodied Simulation: From Neurons to Phenomenal Experience," *Phenomenology and the Cognitive Sciences* 4 (2005): 23–48.

40. Andrew Meltzoff and M. Moore, "Imitation of Facial and Manual Gestures by Human Neonates," *Science* 198 (1977): 75–78; Andrew Meltzoff, "The "Like-Me" Framework for Recognizing and Becoming an Intentional Agent," *Acta Psychologia* 124 (2007): 26–43; Andrew Meltzoff, "'Like me': A Foundation for Social Cognition," *Developmental Science* 10 (2007): 126–34.

41. Meltzoff, "Social Cognition," 27.

42. S. Braten, "Dialogic Mind: The Infant and the Adult in Protoconversation," in *Nature, Cognition and System*, ed. M. Carvallo (Dordrecht, Netherlands: Kluwer Academic Publishers), 1:187–205; S. Braten, "The Virtual Other in Infants' Minds and Social Feelings," in *Dialogical Alternative*, ed. H. Wold (Oslo: Scandinavian University Press, 1992), 77–97; S. Braten, *On Being Moved: From Mirror Neurons to Empathy* (Philadelphia: John Benjamins Publishing Co., 2007), 333; Meltzoff, "Human Neonates," 75–78; Andrew Meltzoff and M. Moore, "Imitation, Memory, and the Representation of Persons," *Infant Behavior and Development* 17 (1994): 83–99; Andrew Meltzoff and R. Brooks, "'Like Me' as a Building Block for Understanding Other Minds: Bodily Acts, Attention, and Intention," in *Intentions and Intentionality: Foundations of Social Cognition*, ed. B. F. Malle et al. (Cambridge, MA: MIT Press, 2001), 171–91; D. Stern, *The Interpersonal World of the Infant* (London: Karnac Books, 1985); C. Trevarthen, "Communication and Cooperation in Early Infancy: A Description of Primary Intersubjectivity," in *Before Speech: The Beginning of Interpersonal Communication*, ed. M. Bullowa (New York: Cambridge University Press, 1979), 321–47; C. Trevarthen, "The Self Born in Intersubjectivity: An Infant Communicating," in *The Perceived Self*, ed. U. Neisser (New York: Cambridge University Press, 1993), 121–73; E. Tronick, "Emotion and Emotional Communication in Infants," *American Psychologist* 44 (1989): 112–19. For a recent, updated survey see V. Reddy, *How Infants Know Minds* (Cambridge, MA: Harvard University Press, 2008).

43. C. Trevarthen, "A Brain for Music," in *Music and the Brain: Current and Future Directions*, ed. P. Robertson and R. Turner (London: Blouin Foundation, 2009).

44. V. Reddy, *How Infants Know Minds*, 41.

45. B. Beebe et al., *Forms of Intersubjectivity in Infant Research and Adult Treatment* (New York: Other Press, 2005).

46. M. Merleau-Ponty, *Phenomenology of Perception* (London: Routledge, 1962), 357.

47. See Ann Cale Kruger, "Imitation, Communion, and Culture," in this volume.

48. See Gallese, "Manifold Nature," 517–28; see also note 32 above.

49. Meltzoff, "Bodily Acts," 171–91; see also Andrew N. Meltzoff, "Out of the Mouths of Babes: Imitation, Gaze, and Intentions in Infant Research—the 'Like Me' Framework," in this volume.

50. Social identification, though, probably falls short of providing a comprehensive account of what it means to be human. Perhaps reciprocity, spelled out in terms of *responsiveness*, is precisely what marks the difference between nonhuman primates and humans. Only among humans do the actions of others imply a *response* rather than a mere *re-action*. It must be added that this dimension of intersubjectivity has been so far rarely explored by cognitive neuroscience.

51. See Martin Buber, *I and Thou* (London: Continuum, 1958).

52. V. Reddy, *Infants Know Minds*.

53. Same as note 39, above.

54. Gallese, "Manifold Nature," 517–28; Gallese, "Shared Manifold Hypothesis," 33–50; Gallese, "Embodied Simulation," 23–48.

55. See A. Goldman, *Simulating Minds: The Philosophy, Psychology and Neuroscience of Mindreading* (Oxford: Oxford University Press, 2006).

56. Ibid.

57. Vittorio Gallese and A. Goldman, "Mirror Neurons and the Simulation Theory of Mind-Reading," *Trends in Cognitive Sciences* 12 (1998): 493–501.

58. Gallese, "Shared Manifold Hypothesis," 33–50; Gallese, "Being Like Me," 101–18; Gallese, "Embodied Simulation," 23–48; Gallese, "Intentional Attunement," 15–24.

59. Gallese, "Before and Below," 659–69; Gallese, "Motor Abstraction," 486–98.

60. Gallese, "Shared Manifold Hypothesis," 33–50; Gallese, "Being Like Me," 101–18; Gallese, "Embodied Simulation," 23–48; Gallese, "Intentional Attunement," 15–24.

61. Edmund Husserl, *Cartesian Meditations* (The Hague: Martinus Nijhoff, 1969); Edmund Husserl, *Ideas Pertaining to a Pure Phenomenology and to a Phenomenological Philosophy*, vol. 2, *Studies in the Phenomenology of Constitution* (Dordrecht: Kluwer Academic Publishers, 1989). Dan Zahavi, "Beyond Empathy: Phenomenological Approaches to Intersubjectivity," *Journal of Consciousness Studies* 8 (2001): 151–67.

62. See M. Jabbi et al., "A Common Anterior Insula Representation of Disgust Observation, Experience and Imagination Shows Divergent Functional Connectivity Pathways," *PLoS ONE* 3, no. 8 (2008): e2939.

63. See Blakemore et al., "Somatosensory Activations during the Observation of Touch and a Case of Vision–Touch Synaesthesia," *Brain* 128 (2005): 1571–83.

64. B. Calvo-Merino et al., "Seeing or Doing? Influence of Visual and Motor Familiarity in Action Observation," *Current Biology* 16, no. 19 (2006): 1905–10.

65. See F. de Vignemont, "The Empathic Brain," 435–41; K. R Stueber, *Rediscovering Empathy: Agency, Folk Psychology, and the Human Sciences* (Cambridge, MA: MIT Press, 2006).

66. In this work, Girard defines the "other" as the *mediator*, which, according to his/her degree of proximity to the subject, is considered either as *external mediator* (when belonging to a different relational domain in space, time, or social status) or as *internal mediator* (when among the subject's peers or close acquaintances). It is the latter, according to Girard, normally perceived by the subject as an obstacle or opponent, to trigger mimetic rivalry; René Girard, *Deceit, Desire, and the Novel: Self and Other in Literary Structure* (Baltimore: Johns Hopkins University Press, 1965), 6, 83.

67. Girard, *Deceit, Desire, and the Novel*, 16.

68. See Pierpaolo Antonello and João de Castro Rocha, "One Long Argument from the Beginning to the End," in *Evolution and Conversion: Dialogues on the Origin of Culture*, René Girard with P. Antonello and J. de Castro Rocha (London: Continuum International Publishing, 2007), 1–16.

69. Gallese, "Shared Manifold Hypothesis," 33–50; Gallese, "Manifold Nature," 517–28.

70. H. De Preester, "From Ego to Alter Ego: Husserl, Merleau-Ponty and a Layered Approach to Intersubjectivity," *Phenomenology and the Cognitive Sciences* 7 (2008): 137.

Part 2

Imitation in Human Evolution, Culture, and Religion

CHAPTER 6

Imitation, Communion, and Culture

Ann Cale Kruger

Despite some physical similarities to our great ape cousins, humans are distinguished by big brains. Brain volume in humans is roughly three times greater than it is in apes.[1] Humans are also distinguished by their ability to create culture—tools, languages, art, institutions, societies, and governments—culture that accumulates modifications over time. Humans transmit their cultural knowledge to subsequent generations, who adopt it and use it as a foundation for cultural innovations that they then pass on. Although nonhuman animal cultures (or proto-cultures) exist, such as seen in chimpanzee termite fishing in the Gombe Stream area,[2] their cultural practices are usually limited by geography and do not progress over the generations. There is no evidence of the "ratchet" effect in nonhuman animal cultures as opposed to human cultures—the accumulation, transmission, and progress of cultural traditions over time.[3] And there are reasons to assert that the differences in generativity between nonhuman and human cultures rest on species differences in teacher and learner transactions.[4]

Is the difference in brain size between apes and humans enough to account for the remarkable difference in the cultures they produce? Can we account for culture by an increase in general intelligence (better memories, faster processing) or by an adapted intelligence (cognitive accommodations to particular environmental pressures)? And which pressures might be the key ones—those pushing for a better understanding of the physical world and thus more impressive amassing of resources, or those that push for cooperative social functioning in complex groups?

In *Things Hidden since the Foundation of the World*, René Girard provides an evolutionary model of how this quantitative physical difference in brain size produced qualitatively different cultural products, describing the "motor for this strange machine."[5] For Girard, brain size increases allowed for complementary increases in the powers of imitation or *mimesis*. Inter-individual rivalry for objects, acquisitive mimesis, was present in ape ancestors, and a source of dangerous conflict as apes reproduced each other's aggressive behavior in competition for something mutually desired. Mimetic powers grew along with brain size in early humans, and the threat mimesis then posed (e.g., reciprocal violence no longer constrained by dominance patterns) initiated the process of hominization, that is, the process of successive adaptations (collective violence, scapegoating, ritual sacrifice) allowing species survival in the context of deadly mimetic violence. Hominization in turn stimulated more complex mimetic systems (rituals, prohibitions, myth) that leveraged symbol use to mediate and control rivalry. In Girard's view, a complete theory of human culture begins with the single principle of mimesis; the complexity of human cultural practices emerged to control its growing and highly destructive potential.[6]

In the years following the development of Girard's expansive and forceful theory, developmental psychologists began to describe with greater precision the mechanisms of imitation in human children. Foundational discoveries, such as the demonstration by Meltzoff and Moore that human infants can imitate facial expressions during the second and third week of life (instead of between the 8th and 12th month as proposed by Piaget),[7] stimulated the growth of theory and research on human social cognition more generally, and especially on those capacities that are available at the time of birth or develop shortly thereafter.[8] These questions about prenatal preparations for human social life and early postnatal development of social cognitive capacities address how human nature reflects and supports cultural evolution. They led to a flourishing of research activity in comparative developmental psychology, producing descriptions of the similarities and differences in how great apes and humans relate to, think about, and learn from others. Through ingenious experimental investigations of ape and human capacities for processes such as joint attention, language, tool use, theory of mind, mirror self-recognition, cooperation, and empathy, Michael Tomasello, Malinda Carpenter, Josep Call, and colleagues have enlarged our understanding and appreciation of the subtle and extensive social cognition of apes, but also have carefully isolated their differences from humans.[9] Their work has led to the identification of a small but powerfully generative difference between ape and human social cognition, a capacity that appears early in human ontogeny and that may explain the emergence, transmission, and refinement of human cultural traditions. Thus, they have produced support for the Cultural Intelligence Hypothesis, the theory that human intellect adapted to function in complex social contexts.[10]

Tomasello, Carpenter, Call, and Moll describe a distinctly human propensity to share psychological states (or intentions) with others.[11] This motivation is present from the earliest weeks of human life and unfolds over the course of the first year to support increasingly sophisticated engagements with others in a "we-centric" space.[12] Sharing psychological states with others is more than, but may include, empathy (a matching of moods), perspective-taking (a shared reference), embodied synchrony (a mirroring of behaviors), theory of mind (an imputation of mental states), or common ground (shared background knowledge).[13] It is having joint thoughts and feelings with another person about some aspect of reality when each is aware of the other's role in the commonality. It is a capacity that I will refer to here as *communion*.

For Tomasello and colleagues, human social cognition follows a developmental pathway that is driven by the infant's emerging concept of person.[14] It arises from our shared primate abilities to understand others' intentions, but is transformed by the human desire to share intimate thoughts and feelings (which is the first definition of "communion" listed in the *Compact Oxford English Dictionary*). It begins with the youngest infants' understanding of other persons as animate actors, and is seen in their dyadic engagement with others for the purpose of sharing.

DYADIC ENGAGEMENT

Human life begins in vulnerability. Altricial, and dependent on the care of at least one willing provider, humans enter the world comparatively unprepared to meet their survival needs. Other primates, and certainly other mammals as infants, are faster to find their own food, make their own way, and stay safe. However, in comparison to other species, the human is born with especially advanced abilities to attend to others and to secure their attention. Babies are born with a suite of preparations for social life that are uniquely advanced, and use them, beginning while yet unborn, to join other humans, to become part of the companionship that is such a vibrant and, in fact, defining characteristic of the species.[15]

Newborns are remarkably organized to see, hear, smell, taste, and learn about others, especially the mother, from the beginning. They prefer the sounds of human language to any other sounds; attend to mother's voice (familiar from uterine experience) and soon thereafter her face, more than to any other; favor the smell and taste of her body and her milk; and coordinate their movements and attention to learn about her as much as they can. From the first minutes of life, their actions and their affect become synchronized with hers. The tuneful, rhythmical interactions between an infant and (usually) her mother start early, develop quickly, and are the foundation for future development.[16]

The contingent and congruent transactions between the infant and caregiver in the earliest days of life are variously referred to as primary intersubjectivity, proto-conversations, interactive synchrony, or affective attunement. They may rely on a basic identification with conspecifics or the understanding of others as "like me."[17] But the focus is on sharing feelings, and the baby's exquisite preparation to engage with others in this way (along with the adult's spontaneous responses to it) secures a place for the baby in the protection and affection of the caregiver. These communications between baby and parent illustrate the early human capacity for dyadic transactions. The shared emotions are facilitated in a number of modalities (through rhythmic movements or cooing and laughing, for example), but a striking feature of these interactions is the mutuality of facial expressiveness, and especially the use of mutual eye gaze, to reach interactive attunement.

Infants look at eyes preferentially over other facial features, and coordinate their eye contact with mothers via their mutual vocalizations.[18] Eye contact serves to initiate or terminate *en face* encounters, and mutual gaze is considered a central component in the formation of attachment. More specifically, young infants are sensitive to the exact direction of another's gaze toward them. As early as five months of age, babies are shown to smile less and attend less when their partner's gaze deviates by as little as 5 degrees.[19] Interestingly, children with autism seem to pay less attention to the eyes of others, and are challenged to detect when others are attempting mutual gaze with them.[20] In situations of complex social-emotional information, high-functioning individuals with autism dramatically differ from typically developing children in their frequent and longer focus on the mouth rather than the eyes, even when no salient information can be found there.[21] Children with autism have significant problems with communication and cooperation, but in more typical development from the earliest ages, the eyes are used to communicate with others a special, mutually understood connection, and this may be foundational for more advanced social and communicative relations.

In chimpanzees (*Pan troglodytes*), infants and mothers also establish communicative repertoires that support the baby's emerging competencies. Their mode of communication is primarily tactile, but Bard, Myowa-Yamakoshi, Tomonaga, Tanaka, Costall, and Matsuzawa show that in some captive settings chimpanzee mother-infant dyads use mutual gaze, and use it more flexibly than in other settings.[22] Bard and colleagues suggest that visual and tactile modalities may be used interchangeably in support of mutual engagement in that species. Although chimpanzees may use eye gaze as one of many means of infant-caregiver co-regulation, the sensitivity to eye gaze and its direction may be particularly human. For example, comparing adult chimpanzees to human infants, Tomasello, Hare, Lehmann, and Call have shown that infants are more sensitive to another's eye gaze direction, while chimpanzees find head direction more salient.[23]

TRIADIC ENGAGEMENT

Around 9 months of age, babies begin to understand other persons as intentional agents who engage in actions to accomplish a goal.[24] When an adult's goal is hard to discern, infants this age often look to the adult's eyes for cues, and check the adult's gaze direction to help sort out what the adult intends.[25] They understand the purposeful nature of adult goal-directed action, and they can discriminate between accidents and persistent trying along the path to success. Infants enter into triadic engagements with adults, sharing goals and perspectives regarding a third item in the world, as when a parent and child play the simple game of rolling a ball back and forth. The child and the adult each know that his or her own and the other's attention is focused jointly, and they think and feel together in the "we-centric" state about an aspect of reality—the ball, and their joint intention to roll it to and fro.

COLLABORATIVE ENGAGEMENT

Shortly after the first birthday, babies understand other people as planners. They understand that an adult can adopt any of a variety of means to reach a goal, and they can discern which plan the adult is pursuing.[26] Babies engage in collaborative engagements with adults creating and enacting a joint plan. By sharing attentions and intentions, infant and adult work with a shared commitment to attain a jointly created goal. In most empirical studies of adult-infant joint attention, the operational definition used is a look to the eyes; the infant knows that the adult knows that they are sharing.[27] For example, some infants this age join adults in a basic turn-taking "conversation" using first words or gestures, as in the "What's that?" pointing and naming game so popular with Western toddlers. The infant points at an object, turns to look in the mother's eyes expectantly, and then turns back to the object as Mother names it, "the cup!" The one-year-old understands that the roles played must be coordinated, and in some activities they are interchangeable. In this way, the enterprise of joint thinking itself becomes an object for joint contemplation. This state of communion becomes part of the concept of the experience itself, and the result is an essentially and deeply social representation of reality. Arguably these are the very skills necessary to acquire, sustain, and advance culture.[28]

COMPARATIVE CONSIDERATIONS

In many ways, the social lives of humans and other primates are similar. Macaques have mirror neurons that resonate to the actions of others,[29] infant chimpanzees imitate facial expressions,[30] and adult chimpanzees emulate the results of another's behavior.[31] As Tomasello and Carpenter have shown, apes will follow another's gaze direction and gesture communicatively to bring about an instrumental goal.[32]

Like humans, other species are motivated socially by both competition and affiliation, and they use their social skills to act on those motivations in appropriate situations. Chimpanzees compete skillfully, create formidable adversarial coalitions, deceive one another to gain advantage, and fight fiercely. They also make peace.[33] In fact, in all species capable of mirror self-recognition (elephants, dolphins, and apes, as well as humans), there are demonstrated acts of compassion for unfortunate individuals, such as targeted helping and consolation, with no obvious direct advantage to the comforter. De Waal argues that the self-other distinction that underlies mirror self-recognition supports these pro-social capacities and constitutes a basic form of empathy.[34]

In many ways, the cognitive lives of humans and apes are also similar. In understanding the physical world, a two-year-old human has no advantage over an ape. Systematic testing comparing young humans (before their cognition could be radically affected by advanced cultural products) and two ape species (adult chimpanzees and orangutans) revealed no differences across the three species in tasks involving understanding or manipulating space, quantities, or causality.[35] Very young human brains have general cognitive capacities equivalent to those of the great apes, suggesting that a general intelligence increase with increases in brain size does not fully explain species differences.

However, Herrmann and colleagues also showed that two-year-old children are dramatically better than apes at tasks recruiting social-cognition skills—theory of mind (including gaze following), communication, and especially social learning (solving a problem by observing a model).[36] As Emery and Clayton put it, apes are good ethologists, but poor psychologists;[37] they understand motivations, perceptions, intentions, goals, and even knowledge of others, but they do not understand that others have mental representations of the world.[38] In tests of theory of mind, no ape has been found to understand false beliefs, which requires a flexible appreciation of another's mental representation of reality. Nearly all children develop this ability spontaneously. Apes will "copy" (emulate) a model's use of a novel tool, usually by reproducing the ends of the demonstrated tool use, but not the means, dropping out or changing steps in the modeled routine that are not necessary to reach the goal.[39] Children not only faithfully copy the means and ends of the model's behavior, but they overimitate, reproducing obviously unimportant and irrelevant details in the model's behavior, possibly because this is a means of learning from adults about causal relations in opaque systems, but certainly not solely to obtain the goal.[40] In their commitment to reproduce every detail of the adult's behaviors, we see evidence of the powerful motivation of children to be with others in a special way—to act and think and feel with them, even if the short-term gain is not evident.

Apes do not attempt to enter into joint attention with others, to intentionally share a common focus for its own sake, as human children do. Only children present a toy to an adult for the sole purpose of sharing attention to it, for the joy of experiencing it together. Only children, as they are about to do a somersault or jump into the pool, shout, "Watch me. Watch me, Mommy. Watch me while I do this!" The urgency with which children request another's attention to an action they perform suggests that the action itself is not satisfactory, or even fully real, unless the other is sharing in it.

This seemingly modest difference between great-ape and human social cognition makes human culture possible. Understanding mental representations, inviting shared experiences, faithfully reproducing the strategies others use—all are the products of deeply social, collective thought. Humans make faithful transmission of culture possible via teaching and learning transactions that literally require seeing eye-to-eye. The human power to create and transmit culture is the result of the synergy of our common primate cognitive capacities and our uniquely human desire to enter into communion with others. By entering into shared states, we are able to learn and transmit the cultural practices and products that are distinctive of the species.

The left sides of figures 1 and 2 represent the cognitive and social-cognitive similarities of humans and apes. They refer to our shared primate heritage of understanding the physical world, understanding the intentions and motivations of others, and our common social emotions fueled by competition and affiliation. The right sides represent the key differences in cultural outcomes between apes and humans. For apes, individuals live in groups characterized by positive and negative encounters, and their group lives are sustained by the

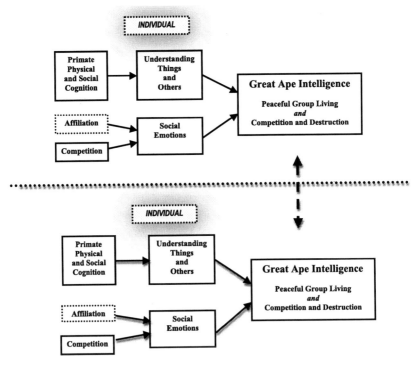

Figure 1.

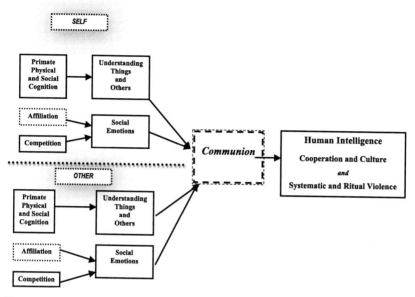

Figure 2.

interactions of their individual capacities. For humans, social life is more than the transactions of individuals; it also involves the intentional communion of self and other to create "we-centric" spaces. Thus, in human life there is the complexity of accumulated and modified cultural traditions, also character-ized by positive and negative enterprises, but encased in intricate systems of meaning. The center of figure 2 illustrates the small but powerful species difference—the sharing of thoughts and feelings, the communion of psycho-logical states—that is the source of culture.

Note that human cultural products, created through the powerful force of communion, may have initially been motivated by competition *or* affilia-tion. In this sense, the theory of hominization proposed here deviates from Girard's, whose emphasis is on the creation of cultural systems (ritual sac-rifice and prohibitions) to control a communal crisis based on competition and uncontrolled mimetic violence. While the practice of ritual sacrifice, as described by Girard, is ultimately a cultural institution of immense commu-nal and cooperative action with affiliative means and ends, its initial impetus is to curb collective violence based on competition. In the model illustrated in figures 1 and 2, culture evolves to solve problems of complex social life, whether motivated by affiliation unrelated to group conflict, or disruptive violence based on competition. Perhaps the main question here is whether or not scapegoating and its subsequent ritualization was *the* communal act out of which all other cultural institutions originated, as hypothesized by Girard. It should be noted, however, that a new data-based theory has been published asserting that cooperation was the primary driving force in the evolution of human intelligence.[41]

Just as the human brain is three times the size of the ape brain, the sclera (white) of the human eye is three times larger as well. This makes detecting the gaze direction of a human very easy, even from some distance. What advantage would it serve humans to "give away" to another person the location of a resource they have spotted, or the identity of an individual with whom they have improperly consorted? Similarly, what is the advantage of human sharing? Very early in life, humans are generous with food and eager to supply information to others. They freely jump into joint adventures with others for no particular reward other than the experience itself. Apes do not share food easily, do not offer information, and do not cooperate with others unless a certain reward awaits them. Their eye anatomy minimizes the sharing of information since their gaze direction is much harder to detect. The Cooperative Eye Hypothesis argues that these anatomical and behavioral differences between the species are markers for the different niches into which they evolved.[42] To live in large and complex social groups, a propensity for helping and sharing was an adaptive advantage for individuals. Humans have survived, even flourished, and been remarkably creative through their fundamental ability to share, to help the other, and thus to create intricate cultural systems to support life. Of course, humans have negative and destructive intentions, but they are often carried out cooperatively (as in armies) and collectively symbolized as sanctified.[43]

Cultural Learning

Michael Tomasello and I distinguished three types of cultural learning that appear over ontogeny (at approximately one, four, and six years, respectively) and that rely on engagement with others with shared intentions.[44] These are descriptions of how an expert and novice, or even two novices, can teach, learn, or create new cultural products through the processes of communion. Cultural learning is different from social learning, in which others may highlight certain aspects of the environment, but the learner is left on her own to figure things out. In cultural learning, the process involves communion, and the product includes a representation of the other person's perspective that is internalized by the learner.

In *imitative learning*, the earliest form of cultural learning, toddlers attend to the intentions of someone performing a task, and then they represent and recreate (sometimes after a delay) the means and the ends of the model. In a series of creative experiments, Andrew Meltzoff has shown that when the model is frustrated or unsuccessful in her attempts at the task, the toddler reproduces what rationally *appeared* to be the model's intended behavior and outcome, not the accidents or false starts.[45] As we have seen, when adults' *intended* behavior is superfluous or unnecessary, children still reproduce it (overimitation).[46] Children are motivated to imitate others not only to reproduce their intentions and outcomes, but also to be in a shared state with them.[47] In addition to the qualitative power of shared states, there is a quantitative relationship between time spent in joint states and the amount learned.[48] For example, compared to singletons, twin toddlers have fewer opportunities to be in shared, one-on-one states with their parents; the frequency of those opportunities positively correlates with their vocabulary size.[49]

In *instructed learning*, the teacher and the learner share the intention of knowledge transmission. The learner has limited understanding, and the teacher has more advanced understanding of the content or procedure to be learned. Each must take steps to enable the learner to reach the teacher's understanding, or to reach a lesser but satisfactory level predetermined by the instructor. To do this, they communicate about their perspectives on the task, and the teacher scaffolds the learner's experience, creating intermediate goals, and monitoring until the learner is successful. They must both work determinedly to coordinate their roles and perspectives and to support the learner to become eventually more like the teacher.

It is worth noting that adults in every culture intentionally instruct their young in this way, focusing on the knowledge or skills they see as being of greatest cultural value.[50] No other species has been observed to instruct their offspring with such intention, and to monitor and adapt the instructional interaction until the learner is successful.[51] Thus, the we-centric teaching and learning described in imitative and instructed learning may explain what makes human cultural transmission possible.

In *collaborative learning*, two individuals begin with comparable levels of knowledge and have the shared goal of solving a problem. They may have differing perspectives on the task, as in the case of two young girls struggling to divide the limited funds they earned in a weekend lemonade-stand venture. The partners describe their current state of understanding, question and critique each other, and ultimately find together a new, more objective understanding. Children are more likely to engage in this critical thinking with their peers than with their parents, and the more they engage in this type of thoughtful exchange, the more they advance in their reasoning.[52] The initial perspectives, the critical stance, and especially the shared purpose of putting their heads together lead the partners to create a new, third perspective. In this way, collaborators create innovations that advance culture.

COMMUNION AND CULTURE IN THE CLASSROOM

Several years ago, I began to wonder whether the theory of cultural learning could be applied to solve practical problems in American education. Is part of the problem of American schools the reliance on rote learning at the expense of shared psychological states between children and those instructing them? In the United States there is an "achievement gap" that begins before the child's first day of school. Children from low-income backgrounds enter kindergarten with less than half the vocabulary of their affluent classmates, placing them at high risk for academic failure.[53] Unfortunately, in schools with large populations of such children, kindergarten instruction often does not resemble child-centered early education, and didactic instruction alone does not suffice where the needed foundational competencies are not in place. In other words, it does not take advantage of the uniquely human forms of learning.

To address these students' need for enriched social and communicative experiences—*cultural learning* experiences—we introduced an intervention in which teaching artists assisted classroom teachers to infuse drama into the language-arts curriculum. From my own experience in theater, I know that drama is inherently mimetic. It calls upon all participants (playwrights, designers, actors, and audience) to identify with each other and to share subjective experiences. In effect, it raises "Watch me while I do this!" to an art form. I reasoned that for children to become encultured in the ways of school, classroom experiences that invite them to share psychological states with others, especially with adults, are necessary.

In our intervention, students, teachers, and artists collaboratively engaged in the mimetic art of story; they created characters, communicated intentions, and made meaning with each other. Sharing one another's lived experiences is essential to creating even the simplest drama, and we hoped this might provide the shared engagements that would support the children's adaptation to the culture of school. (In one of our early studies, an older child said of the playwright-teacher, "Nobody cares about little kids anymore, but Mr. P. listens to us.")

Although the intervention was brief, only 13 lessons over two months, we hypothesized that the emotionally and linguistically rich experience of joint pretense would enhance the children's language development. We anticipated that if students were free to know their teachers and to be known by them in this context—to be in communion—their cultural learning, in this case learning of language, would flourish. During our observation of one class-room visit by a teaching artist, we learned what the possibilities were for children's engagement and identification. As the artist approached, the stu-dents scurried about the classroom searching for the storybook they had used previously as the starting point for their drama. The book was not to be found, but the students kept insisting, "We need the book! Where is the book?" One little girl stepped forward to address her classmates, reassuring them with a gesture that suggested her entire body was opening like a volume of stories, and said, "We don't need the book. We *are* the book." In this proclamation, mimesis and communion are found.

To test our hypothesis, we randomly assigned volunteering schools to intervention and waiting control conditions. Data were collected each year on approximately 100 control and 100 intervention students, 94 percent African American, 71 percent classified as low-income. Before and after the interven-tion, each year for three years, kindergarten students were individually given standardized assessments of language development and a creative writing task. School administrators provided student achievement data.

Kindergarten students in the drama intervention schools showed superior improvement in their writing—the size of the vocabulary they used, the num-ber of sentences, the structure, the theme, and the resolution of their stories all improved compared to controls. They also were significantly more likely to improve in their performance on tests of syntax development than were the students in the control schools. We followed the students as they entered first and second grade to measure any enduring effects of the 13 drama lessons. Without the benefit of any further intervention, the students who were in drama intervention schools in kindergarten had superior report-card grades and superior language-arts achievement-test scores in first grade, and con-tinued to have superior report-card grades in second grade. Students with special needs benefited even more. We have embarked on a four-year study to adapt this intervention to the needs of kindergarten students with limited English proficiency, testing to see if we can enhance their performance in the English-only schools they attend. We expect that the cultural learning process of shared story-making will support the language acquisition of these children to a greater degree that the standard instructional practices do.

Formal education in schools is designed for the transmission of culture. However, large national studies in the United States show that the home environment accounts for most of the variance in students' achievement. Many low-income children enter school without the advantage of cultural consistency between home and school; the supportive processes we associate with first-language acquisition, such as are found in the linguistic environment of the home, must be experienced in the classroom for them to succeed. Creating an environment in school that supports the child's powerful motivation to share feelings and intentions with others will enhance the cultural learning opportunities these children need to become a part of the school community. Communion and cultural learning are the natural process and product of rich human engagement, in the classroom as elsewhere.

CONCLUSION

One of the unique and central features of hominization is the pervasive developmental importance of shared states, or communion. Chimpanzees have impressive capabilities, but they do not have this. As children grow, communion becomes the foundation of cultural learning—first imitative learning, then instructed learning, and finally, the most creative process, collaborative learning. Traditional educational methods, especially for disadvantaged children, have failed to take advantage of this uniquely human developmental course. Recent interventions with drama in disadvantaged schools suggest that supporting these processes can make a significant practical difference in how children learn. The role of mimesis in learning with and through stories has only begun to be explored, and the power and generativity of communion in the classroom, the shared intimacies of thought and feeling, may hold the key to genuine educational reform and would better reflect what we know to be the unique human capacity for cultural learning.

NOTES

1. Harry J. Jerison, *Evolution of the Brain and Intelligence* (New York: Academic Press, 1973).

2. Elizabeth V. Lonsdorf, "Sex Differences in the Development of Termite-Fishing Skills in the Wild Chimpanzees, *Pan troglodytes schweinfurthii*, of Gombe National Park, Tanzania," *Animal Behavior* 70 (2005): 673–83.

3. Michael Tomasello, *The Cultural Origins of Human Cognition* (Cambridge, MA: Harvard University Press, 2000).

4. Christophe Boesch and Michael Tomasello, "Chimpanzee and Human Cultures," *Current Anthropology* 39 (1998): 591–614.

5. René Girard, *Things Hidden since the Foundation of the World* (Stanford, CA: Stanford University Press, 1987), 96.

6. For additional viewpoints on mimesis, see Mark R. Anspach, "Imitation and Violence: Empirical Evidence and the Mimetic Model," and René Girard, "Mimesis and Science: An Interview with René Girard," both in this volume.

7. Andrew Meltzoff and M. Keith Moore, "Imitation of Facial and Manual Gestures by Human Neonates," *Science* 198 (1977): 75–78.

8. See also Andrew N. Meltzoff, "Out of the Mouths of Babes: Imitation, Gaze, and Intentions in Infant Research—the 'Like Me' Framework," this volume.

9. Michael Tomasello et al., "Understanding and Sharing Intentions: The Origins of Cultural Cognition," *Behavioral and Brain Sciences* 28 (2005): 675–735.

10. Esther Herrmann et al., "Humans Have Evolved Specialized Skills of Social Cognition: The Cultural Intelligence Hypothesis," *Science* 317 (2007): 1360–66.

11. Tomasello et al., "Understanding and Sharing Intentions: The Origins of Cultural Cognition."

12. Vittorio Gallese, "The 'Shared Manifold' Hypothesis: From Mirror Neurons to Empathy," *Journal of Consciousness Studies* 8 (2001): 33–50. See also Vittorio Gallese, "The Two Sides of Mimesis: Mimetic Theory, Embodied Simulation, and Social Identification," this volume.

13. Gerald Echterhoff et al., "Shared Reality: Experiencing Commonality with Others' Inner States about the World," *Perspectives on Psychological Science* 4 (2009): 496–521.

14. Tomasello et al., "Understanding and Sharing Intentions."

15. C. Trevarthen, "Intrinsic Motives for Companionship in Understanding: Their Origin, Development, and Significance for Infant Mental Health," *Infant Mental Health Journal* 22 (2001): 95–131.

16. For a summary of infant sociality research, see Philippe Rochat, *The Infant's World* (Cambridge, MA: Harvard University Press, 2001).

17. Andrew Meltzoff, "'Like Me': A Foundation for Social Cognition," *Developmental Science* 10 (2007): 126–34. See also Andrew N. Meltzoff, "Out of the Mouths of Babes: Imitation, Gaze, and Intentions in Infant Research—the 'Like Me' Framework," this volume.

18. D. Stern, "Mother and Infant at Play: The Dyadic Interaction Involving Facial, Vocal, and Gaze Behaviors," in *The Effect of the Infant on Its Caregiver*, ed. M. Lewis and L. A. Rosenblum (New York: Wiley & Sons, 1974), 187–213.

19. S. Hains and D. Muir, "Infant Sensitivity to Adult Eye Direction," *Child Development* 67 (1996): 140–51; L. Symons et al., "Look at Me: Five-Month-Old Infants' Sensitivity to Very Small Deviations in Eye-Gaze during Social Interactions," *Infant Behavior and Development* 21 (1998): 531–36. See also Andrew N. Meltzoff, "Out of the Mouths of Babes: Imitation, Gaze, and Intentions in Infant Research—the 'Like Me" Framework,' this volume.

20. For example, see Jelena Ristic and Alan Kingstone, "Taking Control of Reflexive Social Attention," *Cognition* 94 (2004): B55–B65.

21. Dirk Neumann et al., "Looking You in the Mouth: Abnormal Gaze in Autism Resulting from Impaired Top-Down Modulation of Visual Attention," *SCAN* 1 (2006): 194–202.

22. K. A. Bard et al., "Group Differences in the Mutual Gaze of Chimpanzees (*Pan troglodytes*)," *Developmental Psychology* 41 (2005): 616–24.

23. M. Tomasello et al., "Reliance on Head Versus Eyes in the Gaze Following of Great Apes and Human Infants: The Cooperative Eye Hypothesis," *Journal of Human Evolution* 52 (2007): 314–20.

24. Tomasello et al., "Understanding and Sharing Intentions."

25. C. Moore And P. Dunham Moore, Eds., *Joint Attention: Its Origins and Role in Development* (Hillsdale, NJ: Erlbaum, 1995).

26. Tomasello et al., "Understanding and Sharing Intentions."

27. For example, see Roger Bakeman and Lauren Adamson, "Coordinating Attention to People and Objects in Mother-Infant and Peer-Infant Interactions," *Child Development* 55 (1984): 1278–89.

28. Tomasello et al., "Understanding and Sharing Intentions."

29. Vittorio Gallese and A. Goldman, "Mirror Neurons and the Simulation Theory of Mind-Reading," *Trends in Cognitive Sciences* 2 (1998): 493–501.

30. Masako Myowa-Yamakoshi et al., "Imitation in Neonatal Chimpanzees (*Pan troglodytes*)," *Developmental Science* 7 (2004): 437–42.

31. Claudio Tennie et al., "Push or Pull: Emulation Versus Imitation in Great Apes and Human Children," *Ethology* 112 (2006): 1159–1169; Michael Tomasello et al., "Cultural Learning," *Behavioral and Brain Sciences* 16 (1993): 495–552.

32. Michael Tomasello and Malinda Carpenter, "Shared Intentionality," *Developmental Science* 10 (2007): 121–25.

33. Franz de Waal, *Chimpanzee Politics: Power and Sex among Apes* (London: Jonathan Cape, 1982); Franz de Waal, *Peacemaking among Primates* (Cambridge, MA: Harvard University Press, 1989).

34. Franz de Waal, "Prosocial Primates: Empathy, Fairness, and Cooperation," (paper presented at the Evolution of Brain, Mind and Culture Conference, Emory University, Atlanta, GA, November 12–13, 2009).

35. Herrmann et al., "Humans Have Evolved Specialized Skills of Social Cognition."

36. Ibid.

37. N. J. Emery and N. S. Clayton, "Comparative Social Cognition," *Annual Review of Psychology* 60 (2009): 87–113.

38. J. Call and M. Tomasello, "Does the Chimpanzee Have a Theory of Mind? Thirty Years Later," *Trends in Cognitive Sciences* 12 (2008): 187–92.

39. Michael Tomasello et al., "Cultural Learning," 495–552.

40. Derek Lyons et al., "The Hidden Structure of Overimitation," *PNAS* 104 (2007): 19751–56.

41. Michael Tomasello, *Why We Cooperate* (Cambridge, MA: MIT Press, 2009).

42. M. Tomasello et al., "Reliance on Head Versus Eyes in the Gaze Following of Great Apes and Human Infants."

43. See Melvin Konner, "Sacred Violence, Mimetic Rivalry, and War," this volume.

44. Michael Tomasello et al., "Cultural Learning," 495–552.

45. For example, see Andrew Meltzoff, "Understanding the Intentions of Others: Re-Enactment of Intended Acts by 18-Month-Old Children," *Developmental Psychology* 31 (1995): 838–50. See also Andrew N. Meltzoff, "Out of the Mouths of Babes: Imitation, Gaze, and Intentions in Infant Research—the 'Like Me' Framework," this volume.

46. Lyons et al., "The Hidden Structure of Overimitation."

47. M. Carpenter, "Instrumental, Social, and Shared Goals and Intentions in Imitation," in *Imitation and the Development of the Social Mind: Lessons from Typical Development and Autism*, ed. S. J. Rogers and J. Williams (New York: Guilford, 2006), 48–70.

48. Michael Tomasello and Ann Cale Kruger, "Joint Attention on Actions: Acquiring Verbs in Ostensive and Non-Ostensive Contexts," *Journal of Child Language* 19 (1992): 311–33.

49. Michael Tomasello et al., "Linguistic Environment of 1- to 2-Year-Old Twins," *Developmental Psychology* 22 (1986): 169–76.

50. Ann Cale Kruger and Michael Tomasello, "Transactive Discussions with Peers and Adults," *Developmental Psychology* 22 (1986): 681–85.

51. Michael Tomasello et al., "Cultural Learning," 495–552; Boesch and Tomasello, "Chimpanzee and Human Cultures."

52. Ann Cale Kruger and Michael Tomasello, "Transactive Discussions with Peers and Adults." Ann Cale Kruger, "The Effect of Peer and Adult-Child Transactive Discussions on Moral Reasoning," *Merrill-Palmer Quarterly* 38 (1992): 191–211; Ann Cale Kruger, "Peer Collaboration: Conflict, Cooperation, or Both?" *Social Development* 2 (1993): 165–82.

53. Betty Hart and Todd Risley, *Meaningful Differences in the Everyday Experience of Young American Children* (Baltimore: Paul H. Brookes, 1995).

CHAPTER 7

Imitation and Violence: Empirical Evidence and the Mimetic Model

Mark R. Anspach

FROM IMITATION TO RIVALRY, OR: IF THE OTHER GUY IS SO MUCH "LIKE ME," WHY DOESN'T HE LIKE ME?

Human beings are not solipsistic; they develop their sense of self in interaction with others. This observation is a good starting point for building bridges between empirical imitation researchers and mimetic theorists. Among the contributors to this volume who draw on experimental research, Ann Cale Kruger emphasizes the yearning for communion; Vittorio Gallese speaks of a "we-centric" self that emerges from an intersubjective nexus; Andrew Meltzoff places infant development in the context of a "Like Me–Like You" framework of reciprocal recognition.[1] There is a clear affinity between all these approaches and what René Girard and Jean-Michel Oughourlian refer to as "interdividual" psychology.[2]

The importance of this common ground should not be minimized. On the other hand, a great deal still needs to be done to connect the work of psychologists or neuroscientists on imitation and intersubjectivity to empirical data of the kind collected by Melvin Konner on the prevalence of violence in human (and nonhuman) societies.[3] Here the mimetic theory may hope to make a contribution, for it claims that imitation is a crucial part of the generative mechanism responsible for the prevalence of violence among humans.

This claim marks a radical break with conventional wisdom. Most discussions of conflict and rivalry are still dominated by a "common sense" approach rooted in folk psychology. This approach is *object-centered*; it locates the reason for rivalry in the object at stake.[4] Anything people fight to possess is assumed to be intrinsically valuable—otherwise, why would they want it so much?

Mimetic theory challenges the folk-psychology understanding of rivalry and desire. It replaces the object-centered approach with a "we-centric" one that locates the reason for rivalry in the intersubjective nexus. If a person sees—or merely believes—that another person wants something, he or she is liable to want it, too. The convergence of more than one desire on the same object can be explained by imitation alone, irrespective of the object's intrinsic properties.

Indeed, the folk-psychology assumption that a hotly disputed object must be valuable is *itself based on imitation*: it does no more than reproduce the dynamic in which the perception that other people desire an object makes that object appear desirable. If others are "like me" and they see something as desirable, nothing is more natural than for me to follow their lead. But when I try to get hold of the same thing they want, we may end up fighting. Our very likeness has set us at odds.

The latter claim is counterintuitive. Common sense tells us that avoiding conflict is a matter of "overcoming our differences." If someone else is "like me," shouldn't that make it easier for us to get along? Both Paul Dumouchel and Jean-Michel Ourghoulian show how this assumption can fail in the case of two individuals—for example, two friends who are so alike that they cannot help falling in love with the same woman.[5] Konner cites the Robbers Cave experiment involving two groups of 11-year-old boys who, despite being of like educational, socioeconomic, and religious background, nevertheless developed a fierce rivalry.[6] This experiment clearly demonstrated that rivalry can thrive in the absence of differences, but it did not seek to establish the link between imitation and rivalry posited by mimetic theory.

Indeed, to my knowledge, no experimental studies have yet been undertaken with the express aim of testing claims made by mimetic theory. This lack of attention on the part of empirical researchers may largely be due to unfamiliarity with a theory whose roots are in other disciplines. It is possible, however, that some of the difficulty resides in the all-encompassing nature of the theory itself. So vast is its scope, and so sweeping are its conclusions, that even those who are acquainted with the theory may be unsure how to go about translating it into testable hypotheses. Moreover, an additional difficulty may lie in the novel and sometimes counterintuitive nature of some of the theory's claims. Researchers will only want to investigate these claims if they believe them to be plausible. But how can their plausibility be demonstrated until they are validated by empirical research?

The present chapter represents a preliminary attempt to overcome both of these difficulties. In the first section, I offer an introductory summary of the theory and then reduce its core hypotheses to a schematic model. This model describes a dynamic process, each stage of which hinges upon a specific form of imitation. Breaking the model down into its component parts and identifying more precisely the different types of imitation involved is a necessary first step before the hypotheses can be subjected to empirical investigation. But how can their plausibility be demonstrated in advance?

This difficulty is not as great as it may first appear. If the mimetic theory's view of the world is well founded, evidence in support of its tenets should turn up even in the work of researchers unfamiliar with it. In the rest of the chapter, I shall review some findings from experimental psychology, ethology, and anthropology that tend to confirm the plausibility of various assumptions made in the successive stages of the theoretical model presented in the first section. I shall begin by reanalyzing at some length, from a mimetic perspective, the results of a particularly relevant study pertaining to rivalry in small children. This extended analysis is meant to illustrate the ways in which mimetic reasoning can be applied in an experimental context. I shall then go on to consider how empirical observations on displaced or redirected aggression relate to René Girard's hypotheses on sacrifice and generative violence.

The Theoretical Model: Imitation Every Step of the Way

For René Girard, human violence is not rooted in a biological instinct for aggression, nor in sheer self-centered egotism, but in the very same other-oriented impulse on which human community is founded. Imitation is at the basis of both social harmony and rivalry. It is a more powerful drive in humans than in any other animal because humans are not dominated by instinct. Once their rudimentary material needs are satisfied, they have no way of knowing what they want. The behavior of others is their only guide. Since they cannot find the answer within, they emulate models who seem to have attained an illusory plenitude. At first, a person chosen as a model may be flattered by the attention. But when emulation leads the imitator to want the same thing as the model, the relationship will quickly sour. The person who already has something is unlikely to give it up without a fight. When the model becomes the chief obstacle to the satisfaction of the imitator's desire, admiration is transmuted into resentment and hatred. Thus, the distance from love to war may be shorter than we think. Violence can be the outcome of an interaction that began with warm feelings on all sides. The path to hell is paved with the best of intentions.

The discontinuity between intention and outcome lends violence a certain objective quality, as if it were an external phenomenon independent of human will. Violence may erupt spontaneously, and once it does, it tends to perpetuate itself just as spontaneously. Victims of violence may feel they have no choice but to strike back, even though they well know that the targets of their revenge will be apt to strike back in turn, causing the cycle of revenge to repeat itself with no end in sight. This is a process often observed in regional conflicts today. Thankfully, it is not one most of us have had to confront in our own relatively well-policed communities. In primitive, stateless societies, however, where no judicial apparatus for prosecuting aggressors or settling disputes existed, the most trivial insult could set off an uncontrollable chain of reprisals. If these societies had had no alternative means for dealing with violence, they would have obliterated themselves through internecine conflict. Human culture would have fizzled out before it ever got off the ground.

This, according to Girard, is where religion first came in. Religion gave people a means of coming to terms with their own violence as if it were a transcendent force threatening to overpower them from without. Violence is at the heart of the primitive sacred because it is the primary force inspiring human beings with well-deserved awe and terror. By identifying a fundamental objective reality behind primitive notions of the sacred, Girard has made it possible to explain the universality and longevity that they displayed over most of the course of human existence.

Girard points to two ways in which religion has helped societies manage their internal violence. The first, less spectacular way is through the establishment of classifications and prohibitions that channel desires in different directions and keep rivalries from getting out of hand. Such prudent preventive measures are unlikely to suffice, however. Over time, tensions are bound to build up until they threaten to boil over. The second way in which religion has handled violence is by instituting controlled outlets for it.

These outlets include headhunts, ritualized battles, or holy wars, which redirect violence from within the community toward outsiders, and sacrifice, which directs violence against a neutral victim, such as a prisoner or slave, an animal, or even a plant. No member of the group will be tempted to avenge such a victim, particularly if unanimous participation of all members is required. In short, religiously sanctioned ritual violence is a form of violence that, however bloody it may at times be, leaves no scope for reprisals between members of the group. It is a form of violence that reinforces internal cohesion rather than threatening it, as spontaneous violence ordinarily does.

At this point, a question may arise: how did such providential forms of ritualized violence ever come into being in the first place? Girard's answer to this question constitutes an original solution to the problem of the origins of social order. Unlike social-contract theorists, Girard does not imply that people could discuss things rationally and come to unanimous agreement. If spontaneous violence could be so easily tamed through rational deliberation, it would never have been more than a paper tiger. But if it truly tends to rage out of control, then the answer cannot depend on voluntary human action. A spontaneous process must be envisaged instead. Girard therefore looks for a solution *within the mimetic dynamic of violence itself.*

The solution he finds entails the achievement of unanimity through a displacement of mimetic attention away from divisive objects and towards a common target. In a moment of crisis, when rivalry for the same objects erupts in violence, the objects themselves tend to fade from view as the antagonists shift their attention to one another. Imitation of the rival's desire is replaced by imitation of the rival's violence. The more violence intensifies, the less discriminating it becomes; the more members of the group get caught up in the melee, the more indistinguishable they become. The height of the crisis coincides with the height of undifferentiation. If, for all practical purposes, the antagonists are interchangeable, it is possible for any one of them to become the target of all the others:

> Paradoxically but logically, this intensely conflictual phase, at its most divisive, can become reunitive all of a sudden if it leads not two, not three, not even many, but all the participants without exception to adopt the same enemy through a snowballing of imitation. . . . The paradox of the mimetic cycle is that men can almost never share peacefully an object they all desire, but they can always share an enemy they all hate because they can join together in destroying him, and then no lingering hostilities remain, at least for a while.[7]

The victim may have been singled out arbitrarily, but the unifying effect of the collective murder is real enough, and this explains the supernatural capacity for both evil and good with which the victim will be credited in retrospect by the surviving members of the community. The victim will be scapegoated as the evildoer who must have been magically responsible for the crisis, since getting rid of the victim sufficed to bring the crisis to an end. At the same time, the victim will be worshiped as the benefactor who, by dying, miraculously restored peace and unity to the community.

Subsequent mythology will transfigure the victim into the culture hero from whom all good things came, since internal peace and unity are indispensable to prosperity. Myths, in Girard's reading, are distorted but comprehensible accounts of the crisis and the collective murder through which it was resolved. The distortion derives from a misunderstanding of the actual role of the scapegoat, but behind the fantastic elements, it is possible to make out the type of real event that gave rise to the myth.

As to sacrificial rites, these are attempts to reproduce the benefits that derived from this event by imitating both the original crisis and its resolution. After a while, hostilities will inevitably accumulate once again. The ritualized staging of a unanimous act of collective violence, repeated at periodic intervals, serves to prevent a new crisis from flaring up. Sacrifice is not merely the dutiful commemoration of half-forgotten events involving legendary ancestors; it is the reactivation of a mechanism that is expected to work in the present as it worked in the past—a mechanism whose efficacy is not only symbolic but also real.

This brief summary is necessarily too schematic to do justice to the subtleties of the mimetic theory as it has developed over the last half century. Nonetheless, the essential core of the theory may be fairly presented in schematic fashion. Indeed, what Girard has done is to construct a *model* of human interaction—a model that is deceptively simple, since it is free of all needless complications, but at the same time extremely powerful, since it is capable of generating a multiplicity of complex forms. Rather than disregarding in reductionist fashion the differences that characterize real-life actors and settings, it takes these differences as its object, studying the way they evolve over time and showing how new differences emerge once the highest degree of undifferentiation is reached.

Intrinsically no different from anyone else, the victim is sacralized by virtue of attracting everyone else's violence, and thus becomes the foundation stone upon which a new differential system is edified. "I regard unanimous victimage as a self-regulatory device that can stabilize human communities," Girard remarks, "because it provides a model for the whole elaboration of human culture, beginning with ritual sacrifice."[8] Unanimous victimage is a self-regulatory element of mimetic interaction insofar as it tends to restore the system to balance when it falls into crisis, but it does not rule out the recurrence of new crises. The type of event that Girard posits at the origin of ritual sacrifice is not a once-and-for-all affair, but something that must have taken place again and again in prehistoric times.

It is important to notice, however, that Girard distinguishes *sacrifice* as a ritualized institution from the spontaneous episodes of generative or founding violence that gave rise to the institution.[9] A lynching is not a sacrifice, even though it may provide the template for sacrifice. Girard refers to the victim of an original episode of unanimous victimage as the *surrogate victim*. The surrogate victim is a substitute for all the members of the group. Sacrifice is more complex because it entails a double substitution. The sacrificial victim is not *directly* a substitute for any or even all members of the group; sacrifice does not depend on individual psychological mechanisms, but on a ritual process in which the sacrificial victim is substituted for the surrogate victim, who in turn is a substitute for all the members of the group.[10]

The original, spontaneous substitution was facilitated by the very collapse of differences that characterized the crisis. This is one reason we may speak of a self-regulating mechanism. The more undifferentiated the members of the group become, the easier it is for any one individual to serve as a substitute for all the others. At the height of the crisis, they are all interchangeable doubles. But the individual who ends up as the surrogate victim is transfigured by the role he plays in restoring unity, and thus posthumously becomes the source of a new, differentiated order.

The context in which ritual sacrifice takes place is not the same, because differences still exist and must be maintained and reinforced as a barrier against the spread of violence. The sacrificial victim must therefore satisfy conflicting criteria: it must be similar enough to serve as a recognizable substitute, but not so similar that the killing of the victim too closely resembles the murder of a member of the community, for that would carry the risk of provoking new violence. Ritual victims therefore tend to have marginal status—slaves, children, livestock; they must display both continuity and discontinuity:

> The victim must be neither too familiar to the community nor too foreign to it. . . . The victim should belong both to the inside and the outside of the community. As there is no category that perfectly meets this requirement, any creature chosen for sacrifice must fall short in one or another of the contradictory qualities required of it. . . . In its broadest sense, then, sacrificial preparation employs two very different approaches. The first seeks to make

appear more foreign a victim who is too much a part of the community. The second approach seeks to reintegrate into the community a victim who is too foreign to it.[11]

One might sum this up by saying that the sacrificial victim should be "like me," but not *too much* like me.

The preparation of the sacrificial victim involves a deliberate attempt to *imitate* the surrogate victim, perceived as a "monstrous double," both like and unlike everyone else. More generally, sacrificial ritual is the result of a deliberate attempt to imitate the founding murder. From the emergence of the crisis through its resolution and on to the strategy used to prevent new crises, imitation of one form or another is pivotal every step of the way. In this sense, the mimetic theory is parsimonious in the extreme, for it attempts to derive the entire process from a single principle. Let us conclude by rehearsing the posited sequence in schematic terms:

1. *imitation* of acquisitive desires leads to the choice of the same object, triggering divisive rivalries and violence;
2. the violence intensifies and spreads through *imitation* as each attack provokes reprisals that bring new parties into the fray;
3. once the conflict degenerates into a free-for-all, a snowballing of *imitation* leads to the choice of the same enemy, triggering a unifying and pacifying act of unanimous victimage;
4. subsequent sacrificial rituals are based on the *imitation* of the violent social breakdown and its resolution through unanimous victimage.

Now it is time to see how this model stands up in the face of evidence provided by some empirical studies.

When a Doll Comes between Mother and Child: Anatomy of a Mimetic Triangle

What makes individuals fight over an object? The initial evidence from neurobiology and developmental psychology suggests that the impulse to imitate the acquisitive behavior of others is immediate and pre-reflexive. Gallese and Meltzoff report empirical findings that help us understand why the *hands* and *eyes* of different individuals might tend to converge on the same object.[12] Gallese explains that primates are neurologically "wired" to *reach* in the direction of whatever object another individual reaches for. This predisposition rests on the existence of mirror neurons that "fire both when the monkey performs goal-directed motor acts like grasping objects with the hand and/ or the mouth, and when it observes similar acts performed by others." For his part, Meltzoff demonstrates that human infants *look* in the direction of whatever object another individual looks at. They do not simply mimic an adult's head movement; they follow her gaze. The "inanimate object takes on a special valence when it is looked at by a social other (i.e., it attracts looks of longer duration)." Meltzoff concludes with a discussion of the question: "Is the human infant a natural Girardian?"

In order to take this question further, we would want to study infants' grasping and gaze-following in a context designed to allow for the emergence of rivalry. As Meltzoff notes, the idyllic tête-à-tête between mother and baby is destined to be short-lived:

> The young infant lives in a kind of psychological Garden of Eden. There are two people paying attention to one another in a blissful state of dyadic interaction. This does not last for long. Soon there are interlopers, as the infant becomes aware that third parties are vying for mother's affection. For example, infants begin to pay attention to the fact that mothers do not always look at them, but also cast their gaze to external objects, siblings, and spouses in the environment.[13]

In the rest of this section, I shall examine the results of a particularly suggestive study that tested infants' response to the presence of interlopers.[14] Although this study was not designed with Girardian hypotheses in mind, I will try to tease out as far as possible the implications that it holds for assessing the relevance of the mimetic model.

In experiments conducted by Sybil Hart and her colleagues, a 12-month-old child was made to look on while its mother and another adult woman engaged in lively conversation about either an infant-size doll or an illustrated dessert cookbook. The expectation, confirmed by the experiment, was that the infant would manifest rivalry toward the infant-size doll. This phenomenon had already been described more than a century ago by Charles Darwin, who noted, in a biographical sketch of his 15-month-old son, that "jealousy was plainly exhibited when I fondled a large doll."[15]

Each experimental session was divided into a series of one-minute segments. First, mother and child were quietly seated together in a play area at the center of the room. A minute later, a female stranger entered the room, holding one of the objects, and invited the mother to join her. The two women then sat together on chairs in a corner of the room, and the mother chatted with the stranger about the object while studiously ignoring her child. About 60 seconds into the conversation, the object was switched from the lap of one woman to that of the other. Some 60 seconds later, the stranger exited and the mother again spent a minute sitting quietly with her child in the center of the room. Then the stranger returned and the same scenario was repeated, this time with the other object.[16]

While the infants observed the two adults, the experimenters observed the infants, capturing their reactions on a hidden camera. During each segment of the staged conversation, the infants spent more time looking at whichever of the two adults held the object. When the object was a book, the infants' gaze behavior was object-driven, but when it was a doll, they responded differently depending on whether the doll was held by the stranger or their mother. Both visual gaze toward the mother and proximal contacts with her were greatest when she attended to the doll. The infants showed more distress, as manifested by inhibited play and increased negative vocalizations and protest behavior, whenever the object was a doll, and their distress increased further when the doll was held by their mother.[17]

"Each mother was asked to converse positively about the doll and to handle it as if it were a real infant when it was placed in her lap," the authors note.[18] Clearly, the experimental results are consistent with the prediction that the child would perceive the doll as a rival. Comparing the infants' behavior in the doll situations to the signs of jealousy observed in older toddlers when adults attend to their siblings or peers, Hart et al. emphasize that "infants' disturbances were greater when inattentiveness was demonstrated by their mother, rather than by a stranger, if the object of attention was a doll."[19]

However, the authors of the study also raise the possibility of an alternative interpretation; namely, that the doll, as opposed to the book, was more attractive to the infants because they perceived it as a plaything: "Their protests and proximal contacts with their mother could have reflected frustration with being deprived by their mothers of a preferred toy." Perhaps the only way to settle the question would be to replicate the experiment using a real infant instead of a doll, as Hart et al. suggest. Nevertheless, they provide a convincing counterargument to the alternative hypothesis. If the infants resented being deprived of a toy, one would expect them to make "more bids to play with the doll or extricate it from their mothers. Yet, infants did not contact the doll more than the book." In addition, if the doll's desirability as a toy were the relevant factor, the infants' distal behaviors should not vary depending on who held it: "Since infants' gaze behaviors are not inhibited by strangers and since a stranger is a more novel stimulus than the mother, it is unclear why the increase in stranger-directed gaze was not as great as the increase in mother-directed gaze when given the doll. Why the doll became more compelling as a function of being located on the mother's lap is still unexplained."[20]

Here the mimetic approach could well prove helpful, but we should take care not to apply it too hastily. For example, one might be tempted to conjecture that the doll's desirability in the child's eyes derives mimetically from the attention paid to it by the person holding it. This could lead one to conclude that the doll becomes more compelling when it is on the mother's lap because the mother is a more influential model for the child. But if that were the case, why didn't the book become equally compelling when the mother held it? The mother's influence as a mediator of the child's desire should have proven just as powerful when the book was in her lap.[21]

In reality, by positing the mother as the mediator and the doll as the object, we are going off on the wrong track. We need to come back to the idea that the infant sees the doll as a rival when the mother holds it. In the resulting mimetic triangle, the object of rivalry can only be the mother. The rivalry acts to reinforce *her* desirability in the child's eyes, not the doll's. This conclusion is supported by the fact that she, not the doll, becomes the object of increased attempts at physical contact on the child's part.[22] Consequently, the experimental finding needs to be reformulated: it is not the doll that becomes more compelling as a function of being located on the mother's lap; rather, *the mother herself becomes more compelling by virtue of having an infant-size doll on her lap.*

If the mother is the object in the mimetic triangle, then who is the mediator? The answer should be clear: *the doll is the mediator.* Naturally, I do not intend to suggest that the doll is the source of the infants' love for their mother. She was already the most important person in their lives before the doll arrived on the scene. Yet that is precisely what makes the outcome of the experiment so impressive from a mimetic standpoint. If the mother had acted as the mediator, directing the infant's attention to the doll, the experiment would merely have confirmed the well-known importance of parents as role models. The actual result is much more striking. It provides a vivid illustration of a principle Girard formulates with regard to Proustian desire: "the mediator may be literally *anyone at all.*"[23] In this case, not even a person—a mere inanimate object is what causes the children to suffer sudden pangs of frustrated desire for their mother.

Not only did the infants respond to the doll's proximity to the mother by seeking increased proximity to her themselves, they also displayed more negative or protest behaviors. These reactions seem to have been especially marked when the mother cooed over the doll. "Another intriguing finding of this study," the authors note, "was an association between mothers' positive vocal affect and infants' negative behaviors in the doll condition." This association may appear anomalous: "Typically, more positive affect by mother elicits more positive response from infants."[24] If the infant could be counted on always to imitate the mother, it would respond to cooing with cooing, not protest. Why, in this particular case, does cooing lead to booing—or bawling?

My contention is that the negative response is no less mimetic than the positive one. The difference lies in the identity of the mediator or model. When mother and child are alone together, the infant will tend to take its mother as model and echo her positive vocalizations. In the context of a one-on-one reciprocal interaction, the infant is quite able to see the mother as "like me" despite the gap in age and size. But what happens when the child is left alone in the middle of the floor and watches the mother interact with an infant-size doll? This time, it would seem that the infant perceives a greater likeness between itself and the doll. Yet this does not make the infant *like* the doll any better. On the contrary, it sees the doll as a rival and immediately wants to take its place.

For 60 anguished seconds, the child longs to be where the doll is, happily ensconced in the mother's lap and basking in her positive vocal affect. In short, the child takes the doll as its *model*. One might even say that, at that precise moment, the child wants to *be* the doll: it wants to be the one on the receiving end of the mother's positive vocalizations. And when this desire is thwarted, it may express its frustration through negative vocalizations. These negative vocalizations are in no way imitated from the positive ones, yet they can still be understood mimetically as the result of a desire inspired by a model.

Inspired by a model, but not *imitated* from a model's desire: the doll does not display desire, or indeed manifest any behavior of its own that could be imitated. The mother is the one who puts the doll in her lap. Yet that does not change the fact that, in the resulting triangular relationship with the infant, she is the object of desire while the inanimate object is the mediator. This example should help dispel the misconception, sometimes expressed by Girard's critics, that no mimetic desire could exist in the absence of a "genuine," *non*-mimetic desire to serve as a model.[25] In reality, the attribution of superior "authenticity" to the model is itself an illusion symptomatic of mimetic desire.

Girard cites the narrator of *Swann's Way*, who recalls having felt that everyone around him was "endowed with a more real existence."[26] This haunting sense of inward inadequacy lends mimetic desire what Girard calls its "metaphysical" character—a description that is potentially misleading insofar as it conjures up the type of existential anguish associated with intellectuals smoking Gauloises in Parisian cafés. Surely 12-month-old babies lack the philosophical sophistication to suffer from such rarified emotions?[27]

The results of the doll experiment put these considerations in a new light. When the mother studiously ignores her baby and handles the doll as if *it* were the real infant, the baby reacts as if the doll were truly "endowed with a more real existence," thus demonstrating that no philosophical sophistication is necessary to suffer from anguished sensations of inward inadequacy. Indeed, such feelings are probably the common lot of human babies, whose radical lack of autonomy frequently reduces them to weeping and wailing. Conversely, the most acute forms of existential anguish in adults may result from a breakdown in the elaborate psychic scaffolding that separates grownups from the type of desperate helplessness displayed by the infants in the experiment.

Faced with the sudden appearance of a rival for their mother's attention, the infants manifested a wide range of more or less anguished responses. In addition to producing the aforementioned negative vocalizations, they "demonstrated disorganized behaviors (such as stilling), evidence of anxiety (such as rocking, pacing, self-clinging, propitiatory smiling and numerous self-comforting and avoidance responses) as well as evidence of intense distress (such as panic cries, intense wailing, temper tantrums and aggression toward the mother or the doll)."[28]

Culminating as it does with temper tantrums and outright aggression, the enumeration of the infants' responses to the doll situation reminds us of how easily rivalry may lead to violence. This brings us to the next stage in our schematic presentation of the mimetic model. Although the authors of the study say nothing further about the acts of aggression they witnessed, it seems to me that two significant facts are concealed in the simple observation that the targets were "the mother or the doll."

First, it is significant that *only* the mother and the doll were targets of aggression. After all, the experimental situation included another possible target: the female stranger with whom the mothers chatted about the doll. One might have expected the children to resent this other woman as a flesh-and-blood rival for the mother's attention, yet whatever negative reactions the stranger aroused do not seem to have spilled over into outright aggression. Why not? The answer, no doubt, is that an infant does not perceive a grown woman to be as much "like me" as an infant-size doll. The doll's greater resemblance to the child intensifies the rivalry. In short, as Girard would predict, *the risk of violence rises as differences diminish.*

Second, it is also significant that *both* the mother and the doll were targets of aggression. This result was not necessarily predictable. The infants might have directed their aggression always at the same target. Instead, they sometimes lashed out at the rival—the doll—and sometimes at the very object they feared losing: the mother. To be sure, neither reaction is surprising on the part of a person in the grip of jealousy, and either could be construed as rational: striking at the doll could be a means of keeping it away from the mother; striking at the mother could be a means of punishing her for cradling the doll. Or perhaps such rational considerations are superfluous when it comes to explaining why a baby would direct aggression toward a person or thing that has aroused its anger. Whatever the case may be, the fact remains that *the direction the aggression takes is underdetermined.* This opens up the theoretical possibility that aggression and violence can be *redirected* from one target to another, as Girard contends. In the next section, we shall examine several forms of empirical evidence that support this key tenet of the mimetic model.

Redirected Aggression, Imitation, and Reciprocity: Kicking the Dog, Biting the Master

In the opening chapter of *Violence and the Sacred*, René Girard emphasizes what might be called the *mobility* of violence—its tendency both to propagate from one actor to another and to shift its attention from one target to another. This mobility is what makes it so difficult to contain. Once violence erupts, it threatens to spread out of control. Yet the same mobility that renders violence perilously contagious also lends it a ductile quality, making it susceptible to being channeled in less dangerous directions. This, Girard contends, is where sacrificial rituals come in. Exploiting "the ability of violence to move from one object to another," these rituals "serve to polarize the community's aggressive impulses and redirect them" toward expendable substitute victims.[29]

Is this understanding of sacrificial ritual plausible? While marshaling an impressive array of anthropological evidence in support of his theory, Girard spends little time substantiating the underlying assumption that aggressive impulses can be shifted from one object to another. Nevertheless, a large and growing body of empirical research on redirected or displaced aggression leaves little doubt as to the reality of the phenomenon.

The notion of displaced aggression has a long history in psychology, going back at least to the 1939 publication of *Frustration and Aggression* by Dollard et al.[30] According to the frustration-aggression hypothesis, anger that cannot be directed toward a given source of frustration may end up being displaced onto an external target. Thus, when the level of aggression manifested toward a target seems incommensurate with the level of provocation supplied by that target, the reason may lie in the subject's failure to respond aggressively to an earlier provocation. The classic example is that of the employee who restrains himself from answering a rebuke from his boss only to go home and kick the dog for barking a friendly greeting at him.[31]

While the "kick-the-dog effect" has entered popular parlance, a content analysis of 122 social psychology textbooks performed by Amy Marcus-Newhall et al. reveals that the concept of displaced aggression fell out of academic fashion after an initial surge of interest following the publication of the seminal work by Dollard and his colleagues.[32] Even Girard has shown himself wary of appearing to attach excessive importance to a phenomenon sometimes deemed marginal. "Clearly," he writes, "it would be inexact to compare the sacrificial act to the spontaneous gesture of the man who kicks his dog because he dare not kick his wife or boss." And yet, he adds, "there are Greek myths that are hardly more than colossal variants of such gestures," such as that of Medea, who, in the absence of the faithless husband who has provoked her rage, directs her violence against her own children.[33]

The gestures of Medea or the aggrieved employee possess a spontaneity that is alien to ritual sacrifice. Nonetheless, if Girard is right, the latter also embodies displaced aggression. What, then, is the empirical status of the phenomenon? Are social psychology textbooks justified in neglecting displaced aggression? According to Marcus-Newhall et al., the answer is an emphatic no. They conducted a comprehensive meta-analysis of the experimental literature and found that displaced aggression constitutes a robust effect: "The obtained mean effect size of +0.54 (with a 95% confidence interval that very clearly does not include zero) shows that those who are provoked and unable to retaliate reliably respond more aggressively toward an innocent other than those not previously provoked."[34]

It is worth taking a moment to look more closely at why a subject should prove unable to retaliate against the source of the original provocation. Typically, three types of situations exist, two of which we have already encountered: the subject is afraid to retaliate against the provoking agent (the employee and his boss),[35] or else the provoking agent is out of reach (Medea and her absent husband). The third type of situation is a little different. It occurs when the provoking agent is not another person, but something intangible, such as oppressively hot weather, against which it is impossible to strike back directly.[36] At first glance, this type of situation may appear to depart from the mimetic schema, where rivalry among individuals is the point of departure. In reality, however, the existence of an intangible provoking agent is characteristic of the periods of crisis in which Girard says conflicts come to a head, triggering the search for a scapegoat. Plagues provide the backdrop for the expulsion of Oedipus in Sophocles' tragedy or the real-life persecution of Jews in the Middle Ages. Simon Simonse has studied how a prolonged lack of rain can lead to the execution of the king or queen in the Sudan.[37] As for oppressively hot weather, it figured prominently in Spike Lee's 1989 movie *Do the Right Thing*, which dramatized the way tempers can flare when temperatures rise, exacerbating already latent ethnic tensions.

The economic climate is also an intangible provoking agent against which individuals have no recourse. A recent reexamination of the evidence on lynchings of African Americans in the South between 1882 and 1930 found that their frequency increased with declines in the price of cotton, thus confirming the classic sociological hypothesis that "mob violence against southern blacks responded to economic conditions affecting the financial fortunes of southern whites."[38] Finally, military setbacks or loss of territory can also act as intangible provoking agents, prompting national governments to vent their ferocity on scapegoat populations. Such circumstances can be shown to have played a key role in three genocides of the twentieth century: those of the Armenians of Anatolia, the European Jews, and the Rwandan Tutsis.[39]

Redirected aggression is observable in animals as well as people. A moment ago, we spoke of the "kick-the-dog" effect in which a man returning from a bad day at the office takes his anger out on the family pet. Dog owners may be familiar with a similar phenomenon, which I propose to call the "bite-the-master" effect. This form of redirected aggression occurs when man's best friend turns on an owner who stops him from fighting with another animal. According to Debra F. Horwitz, "If during the course of a dogfight, one owner picks up an animal, the other animal may continue to attack and direct that attack at them."[40] The *Merck Veterinary Manual* defines *redirected aggression* as "aggression that is consistently directed toward a third party when the dog is thwarted or interrupted from exhibiting aggressive behaviors to its primary target," and notes that the same behavior is also seen in cats.[41]

The dynamic of the "bite-the-master" effect differs somewhat from that of the "kick-the-dog" effect. In the former, the same dog is the source of both the initial aggression and the redirected aggression. When a hopping-mad

Spot is deprived of its primary target, Fido, it redirects its violence against a third party, the hapless human owner. The new attack is, as Horwitz suggests, a *continuation* of the original attack. Only the target has changed. The "bite-the-master" effect demonstrates the ease with which violence can be diverted toward a substitute victim.

The tendency of violence to shift its attention from one *target* to another is one side of what I defined, at the beginning of this section, as the *mobility* of violence. I identified the other side of this mobility as the tendency of violence to propagate from one *actor* to another. The "kick-the-dog" effect illustrates this second tendency. Here, the original source of aggression is the boss who gives his employee a tongue-lashing, while the source of the redirected aggression is the employee who, afraid to lash back at his boss, lashes out at his dog instead. This time, the violence—verbal or physical—propagates from one actor to another. It is still possible to speak of violence being diverted toward a substitute victim, assuming that the employee's first preference would have been to kick his boss; but the boss in this case is only a potential victim, not an actual one.

When the aggrieved employee kicks his dog, the new attack is no longer a continuation of the initial attack, at least not directly so, yet there is a sense in which the original violence has been transmitted or passed along. It may be useful here to think about the transmission of violence in terms of *reciprocity*. Anthropologists recognize two forms of reciprocity, direct and indirect. In *direct* reciprocity, A gives something to B, and B responds by giving to A. If the employee were to give his boss a tongue-lashing in return for the tongue-lashing received, that would be direct reciprocity. If, on the other hand, he holds his tongue only to give his dog a kick later, that would be *indirect* reciprocity. In the latter type of reciprocity, A gives something to B, and B responds by giving to C.

Indirect reciprocity is also called *generalized* reciprocity because, being open-ended, it can produce a continuous chain that spreads from one individual to another like cascading dominoes, with C giving to D, D giving to E, and so on. If the dog who receives a kick is afraid to snap back at his angry master, he may turn around and bite the cat, thus continuing the chain. For this reason, redirected aggression is sometimes also called *cascaded* aggression. Peter C. Reynolds points to a "great morphological similarity between redirected aggression and the chaining of play chases that occurs in young rhesus monkeys. In play-chase chaining, which the author and fellow observers dubbed *billiard ball* playing, the terminator of one chase goes on to begin another, and the encounters bounce around the social group from one monkey to the next."[42] Play-chase chaining is a good example of generalized reciprocity.

From a mimetic point of view, both types of reciprocity can be understood as having their basis in imitation: whatever A does, B does likewise. At the same time, the existence of the two different types of reciprocity highlights a fundamental ambiguity in the concept of imitation: if A does something to B, we may predict that B will be inclined to imitate A's action, but what does that mean in practice? Will B imitate A by doing the same thing back to A, or will he do to a third party what A did to him?

This question is not relevant only in the case of violence. It arises whenever the concept of imitation is used in an interactional context. Let's say that B is standing in a field to the north of A. A throws a ball to B and B catches it. Now B is going to imitate A's action. Will he do so by lobbing the ball back to A, or will he turn around and toss it to C, who is standing to the north of him? Merely to say that B will imitate A does not give us enough information to answer the question. This is a problem that has not received sufficient attention from either students of mimetic theory or imitation researchers.

Michael Tomasello and Malinda Carpenter raise an analogous problem in their discussion of imitation of reciprocal behavior, which they treat as a special case in the context of children's learning from adults: "There are some kinds of actions that children observe and attempt to imitate that have a special structure because they involve people having goals toward one another reciprocally."[43] Tomasello and Carpenter give the example of a role-reversal game in which a mother "blows a raspberry" on her child's arm and the child must learn to "blow a raspberry back on her mother's arm." Where we have been comparing two forms of reciprocity, direct and indirect, Tomasello and Carpenter compare direct reciprocity to no reciprocity. They note that when the mother blows a raspberry on the child's arm, the child could very well imitate her mother's action without reciprocating: "She might blow a raspberry on her own arm, in exactly the same place the mother did."

One could also imagine an indirectly reciprocal response. For example, if the child possesses a doll, she might imitate her mother's behavior by blowing a raspberry on the doll's arm. I propose, therefore, that imitation can in fact take at least three forms: non-reciprocal behavior, directly reciprocal behavior, and indirectly reciprocal behavior. While indirect or generalized reciprocity does not involve role reversal, it would nevertheless seem to entail a higher level of complexity inasmuch as it requires the imitator to generalize beyond the interaction with the initial partner by using the same behavior in a relationship with a third party. This is particularly evident in the case of violence, where striking back at the person who hit you is a simpler and more natural reaction than turning around and walloping an innocent bystander.

But it is also a more natural reaction than directing a blow at oneself. In the case of violence, at least, imitation is much more likely to assume a directly reciprocal form than a non-reciprocal form. If an aggressive playmate hits a child in the arm, there is little chance the child will imitate this action by trying to hit his own arm in exactly the same place. Instead, he will direct his blow at the other child without stopping to think about it. Where violence is concerned, role reversal comes naturally. There is no need to *learn* to behave toward the other person the way the other person behaved toward oneself.

In this sense, reciprocal violence differs from the kinds of action studied by Tomasello and Carpenter or other imitation theorists that involve people having goals toward one another reciprocally. The mother who blows a raspberry on the child's arm wants the child to blow a raspberry on her own arm, and the child must grasp this reciprocal structure for successful imitation to take place. On the other hand, the boy who hits his playmate probably does not want the playmate to hit him back, yet this is unlikely to stop imitation from taking place. And even if the two boys do end up having goals toward one another reciprocally—i.e., each wants to clobber the other—this symmetrical structure emerges spontaneously, not as the result of intentional coordination.

According to Tomasello and Carpenter, "Children reproduce what they understand others to be doing, and so their social learning and imitation depend crucially on their skills of social cognition, especially those involved in reading the intentions of other people."[44] Although this might be true in general, I would argue that violence could well constitute an exception to the rule. The impulse to strike back when physically attacked probably does not depend on intention reading. From an evolutionary point of view, there should be a clear survival value in the ability to launch an immediate counterattack without stopping to think about an aggressor's intentions. This survival advantage underlies the philosophy of the trigger-happy cop: "Shoot first and ask questions later."

Of course, such a philosophy is rightly repellent to enlightened thinking. It is a throwback to a more primitive stage of evolution.[45] Human culture requires that individuals learn not to be too fast with the trigger. Otherwise, the group will succumb to infighting among its members. Humans cannot respond reflexively to every perceived aggression with physical violence; they need to stop and think about what the other party's true intentions are. In this crucial respect, intention reading is necessary for group survival. Indeed, its role in the imitation of socially desirable behaviors may be no more than an evolutionarily fruitful side benefit. One might conjecture that the most basic

function of intention reading is to *inhibit* unnecessary and potentially dangerous imitation. Knowing when not to duplicate the behavior of others is clearly
an indispensable part of social cognition. Thus, children must learn that if
someone jostles them, they should weigh the other party's intention before
they reproduce the action by jostling in return. If shoving matches broke out
every time one person bumped accidentally into another, riding a crowded
elevator or subway car would be impossible. Civilization rests on overcoming
the natural propensity to strike back.

The entire theory of displaced aggression is based on the premise that
striking back is indeed the "normal" or default reaction. As Marcus-Newhall
et al. observe, "When an individual is provoked, direct retaliation toward
the original provocateur becomes the goal state."[46] This happens whether
or not the provoked individual understands the role reversal involved. The
existence of reciprocal goals between participants in a slugfest is not the result
of conscious imitation. Perhaps for this reason, aggression theorists tend not
to speak of imitation at all. Conversely, to the extent that imitation theorists
are preoccupied with cognitive issues pertaining to learning, they tend not to
talk about violence. The originality of the mimetic theory lies in the attempt
to connect imitation and violence in a single explanatory model. This model
aims to show how *non-conscious* imitation can lead to both violence and reconciliation through the unfolding of a spontaneous dynamic.

The pivotal moment in this dynamic comes when the divisive violence
of all against all gives way to the unifying violence of all against one. In a
word, all the violence is simultaneously *redirected* in a single direction. The
myriad conflicts that were tearing the community apart are suddenly resolved
when everyone shares in lynching the same victim. So dramatic is the shift
from violence to peace that it appears downright miraculous. Indeed, if René
Girard is right, this seemingly providential turn of events is what first gives
rise to the perception of divine intervention in human affairs. Subsequent
sacrificial rituals are deliberate attempts to reproduce as faithfully as possible—to *imitate*—the spontaneous act of generative violence through which
a community at war with itself unwittingly managed to restore peace.

This account of generative violence inevitably remains the most controversial aspect of the mimetic theory because it claims to describe a type of
event that can never be observed. The same is not true of the other links in the
chain of reasoning. The convergence of desires on a single object, the emergence of violent rivalries to possess the object, the perpetuation of violence in
self-perpetuating cycles of revenge, the collective killing of a scapegoat by a
lynch mob, the collective killing of a sacrificial victim in a religious ritual: these

are all well-known, observable phenomena that Girard has brought together for the first time in a single unified theory. One may still dispute the way the theory explains these phenomena, but not whether they exist.

The case of generative violence is different. Here the theory posits the existence of a type of event, or sequence of events, that could only have taken place long ago, in prehistoric eras before the establishment of any religious system, or in extreme moments of "sacrificial crisis" when the earliest ritual systems broke down and all civilized brakes on the imitation of violence collapsed. One might jump to the conclusion that even if events of this kind occurred, all memory of them must be lost in the mists of time. What Girard contends, on the contrary, is that memories of such events have been preserved, albeit in distorted form, in myths and rituals found throughout the world.[47]

In particular, Girard has drawn attention to the fact that sacrificial rituals often include a preparatory phase characterized by reciprocal violence, suggesting that "the ritualistic imitation deals first with the sacrificial crisis itself, with the chaotic antecedents to the unanimous resolution." In the Greek Bouphonia, for example, "the participants make a point of quarreling among themselves before turning their attention to the designated victim." Indeed, Girard writes, "All the mock battles that generally take place prior to sacrificial ceremonies and all the ritual dances whose formal symmetry is reflected in a perpetual confrontation between the performers" can be interpreted as imitative reenactments of a crisis that was resolved through the "metamorphosis of reciprocal violence into unilateral violence" against a single victim.[48]

How could such a dramatic metamorphosis originally have taken place? It is one thing to reenact successively the reciprocal and unilateral phases of violence in the course of a ceremony; it is another to imagine how such a sudden shift could occur spontaneously in the heat of fighting. As we saw earlier, Girard emphasizes the way in which the snowballing of imitation could lead more and more individuals to pile onto the same target. The greater the number of those attacking a given person is, the more powerful will be the attractive force they exert, until ultimately everyone joins in and unanimity is achieved. The surrogate victim ends up as the substitute for all the other antagonists, whom the confusion of the melee has already turned into interchangeable doubles.

While the logic of a progressive convergence on unanimity is clear enough, one might still ask how readily individuals who are caught up in fighting among themselves will turn their violence against someone else. In fact, what is demanded of each individual antagonist is that he *redirect* his violence from an initial target to a third party. Can our earlier discussion of redirected aggression and reciprocity shed any light on this problem?

I began by distinguishing the "kick-the-dog" effect from what I called the "bite-the-master" effect. It seems to me that—at least as a first approximation—the scenario of generative violence can be reasonably assumed to combine both. The "bite-the-master" effect demonstrates that there is nothing implausible in the idea of a creature in the heat of battle shifting his attack from one target to another. Nor does this effect necessarily depend on the intervention of a "master" able to step in and deprive an assailant of his original target. The latter could flee on his own, evade a blow, or simply be knocked to the ground, prompting the assailant to continue his attack against a substitute victim. As to the person who got away, rather than returning to strike back at the original attacker, he might "kick the dog" by directing his violence against a less-formidable substitute—and, guided by imitation, he might even choose the *same* substitute. In this way, the "kick-the-dog" and "bite-the-master" effects would converge, leading to the same result. Indeed, to the extent that the initial violence is truly reciprocal, with each antagonist giving as good as he gets, it may be difficult in practice to distinguish attackers and attackees. That is one reason they qualify as interchangeable doubles in the first place.

Everyone is equally responsible for the violence, yet everyone perceives himself as a victim of it. From the vantage point of any one individual, the total sum of violence raging around him will be far greater than his own personal contribution to it. In this sense, each antagonist is genuinely a victim of events beyond his control. I would argue, therefore, that *the victim's role must be taken to be primary.* This is consistent with Girard's observation that the persecutors of a scapegoat always view themselves as the victims. They lash out at the scapegoat as a stand-in for an oppressive situation against which they have no recourse, just as the employee who feels victimized by an oppressive boss might lash out at his pet. It follows that the underlying logic is that of "kicking the dog," except that here the lashing out is not deferred; it occurs in the heat of battle with the immediacy of "biting the master."

In our earlier discussion of "kicking the dog," I introduced a distinction between two types of reciprocity, direct and indirect. I went on to propose that imitation can take the form of behavior that is directly reciprocal or indirectly reciprocal. In the initial phase of the crisis, the violence is directly reciprocal: each antagonist imitates the person who hit him by hitting back, as if the two parties were playing an ongoing role-reversal game. How, then, is the shift effected to unilateral violence against a third party? Girard refers to this shift as a veritable "metamorphosis." It is my contention that this metamorphosis rests on a shift in the *form of imitation*. There is nothing particularly mysterious or uncanny about it. The same shift occurs whenever direct reciprocity gives way to indirect reciprocity.

Imagine a father and son alone in the yard. The father tosses a ball to his son; the son imitates him by tossing it back: *direct reciprocity*. The father again tosses the ball to his son. Just then the boy's brother arrives. This time, rather than tossing the ball back, the boy turns around and throws it to his brother: *indirect reciprocity*. The boy is still imitating his father, but he is doing so by behaving toward his brother the way his father behaved toward him. Nothing has changed but the direction of the imitation.

It may seem counterintuitive to speak of indirect reciprocity in the case of unanimous violence against a surrogate victim. After all, indirect reciprocity is generally open-ended; it is associated with continuing chain reactions, whereas Girard tells us that unanimous victimage closes off all possibility of further reprisals. That is true enough, but it is unanimity that does the trick. Once everyone has participated in the collective murder, no one is left to avenge the victim. There is nobody else in the field who can pick up the ball and run with it. Nevertheless, the apparent metamorphosis conceals a switch between direct and indirect forms of reciprocity, rather as if each individual felt victimized by someone else and, instead of fighting back, turned around and lashed out at a dog: reconciliation will be achieved provided that all the antagonists choose to kick the *same* dog.

Or the same pig. I want to conclude with an account of a pig sacrifice performed to end a vendetta between hostile groups in highland New Guinea. Such peace ceremonies constitute an important form of evidence that Girard himself never studied. These rituals allow us to observe first-hand what human groups do when they wish to escape from the reciprocal violence of a blood feud.[49] As the mimetic theory would predict, reconciliation is achieved through sacrifice. The last group that killed a member of the rival group must make the peace overture. If it wants to keep the other group from coming to kill one of its own members, it must offer one or more animals as substitute victims. After the sacrifice, the two groups inaugurate peaceful relations by sharing together in a common meal.

A Daribi peace ceremony described by Pierre Lemonnier makes the underlying logic of such rituals unmistakably clear.[50] In New Guinea, pork is the most appreciated festive dish. To make peace after a murder, the killer's group must bring pigs to the victim's group. As the emissaries approach leading their pigs, the intended recipients raise their bows to take aim at the advancing men. Revenge is in easy reach. If they release their bowstrings, they can do to their enemies what their enemies did to them: *direct reciprocity*. But they know their joy would be short-lived: the roles would soon be reversed when their enemies came back to take vengeance in turn. That is why they have chosen to make peace—not by renouncing violence outright, but rather by *redirecting* it toward a safer target. After a moment of hesitation, they slowly turn their bows toward the animals: *indirect reciprocity*. They will imitate what their enemies did to them by doing the same thing—to the pigs.

ACKNOWLEDGMENTS

I am grateful to Scott Garrels for his help in the course of writing this chapter, to Jean-Pierre Dupuy and Lucien Scubla for useful comments on the completed manuscript, and to Imitatio for providing the financial support necessary for me to carry out the research. The responsibility for the ideas expressed is mine.

NOTES

1. See Ann Cale Kruger, "Imitation, Communion, and Culture," Vittorio Gallese, "The Two Sides of Mimesis: Mimetic Theory, Embodied Simulation, and Social Identification," and Andrew N. Meltzoff, "Out of the Mouths of Babes: Imitation, Gaze, and Intentions in Infant Research—the 'Like Me' Framework," all in this volume.

2. See René Girard, *Things Hidden Since the Foundation of the World*, trans. Stephen Bann and Michael Metteer (Stanford, CA: Stanford University Press, 1987), book 3, and Jean-Michel Oughourlian, "From Universal Mimesis to the Self Formed by Desire," in this volume.

3. See Melvin Konner "Sacred Violence, Mimetic Rivalry, and War," this volume.

4. For a critical discussion of the "classical ontology" holding that "discord arises over the possession of things," see Roberto Farneti, "A Mimetic Perspective on Conflict Resolution," *Polity* 41, no. 4 (October 2009).

5. See Jean-Michel Oughourlian, "From Universal Mimesis to the Self Formed by Desire," and Paul Dumouchel, "Emotions and Mimesis," in this volume.

6. See Melvin Konner "Sacred Violence, Mimetic Rivalry, and War," this volume.

7. These remarks of Girard's are excerpted from the transcript of a discussion of his theory included in Robert G. Hamerton-Kelly, ed., *Violent Origins: Walter Burkert, René Girard, and Jonathan Z. Smith on Ritual Killing and Cultural Formation* (Stanford, CA: Stanford University Press, 1987), 126, 128.

8. Hamerton-Kelly, *Violent Origins*, 121.

9. The term "founding violence" is potentially misleading because it could suggest that the event already possessed an institutional character. Unlike "foundation sacrifices," truly founding violence in Girard's sense is not carried out with the deliberate intent of founding anything; it is only in retrospect that it may be recognized as having unwittingly laid the foundation for what came later.

10. René Girard, *Violence and the Sacred*, trans. Patrick Gregory (Baltimore: Johns Hopkins University Press, 1977), 101–2; cf. 269.

11. Girard, *Violence and the Sacred*, 271–72.

12. See Vittorio Gallese, "The Two Sides of Mimesis: Mimetic Theory, Embodied Simulation, and Social Identification," and Andrew Meltzoff, "Out of the Mouths of Babes: Imitation, Gaze, and Intentions in Infant Research—the 'Like Me' Framework," in this volume.

13. Andrew Meltzoff, "Out of the Mouths of Babes: Imitation, Gaze, and Intentions in Infant Research—the 'Like Me' Framework," in this volume.

14. Sybil Hart, Tiffany Field, Claudia Del Valle, and Marc Letourneau, "Infants Protest Their Mothers' Attending to an Infant-Size Doll," *Social Development* 7, no. 1 (1998): 54–61. I am grateful to William Hurlbut for bringing this study to my attention.

15. Charles Darwin, quoted in Hart et al., "Infants Protest," 60.

16. Hart et al., "Infants Protest," 56.

17. Ibid., 56–57.

18. Ibid., 56.

19. Ibid., 59–60.

20. Ibid., 59.

21. The mean proportion of time the infants gazed at the mother when she held the book (33.8) was only slightly higher than the time they gazed at the stranger when the latter held the book (30.6). By contrast, when the mother held the doll, the corresponding figure (63.0) was about 50 percent higher than when the stranger held the doll (41.1) and nearly double what it was when the mother held the book (see Hart et al., table 2, 59).

22. See Hart et al., table 2, 59: When the mother held the doll, the mean proportion of time the infants touched the *object* (13.5) was about the same as when she held the book (13.3)—and little more than what it was when the stranger held the book (10.8)—but the time spent touching the *mother* rose to 52.2 from only 16.4 when she held the book and 27.0 when the stranger held the doll.

23. René Girard, *Deceit, Desire, and the Novel: Self and Other in Literary Structure*, trans. Yvonne Freccero (Baltimore: Johns Hopkins University Press, 1965), 92.

24. See Hart et al., 60. The authors specify that "the covariate, vocalizations by the mother, was not significant alone."

25. See Jean-Pierre Dupuy, "Naturalizing Mimetic Theory," this volume, where he discusses the case in which two mimetic desires serve as models for each other.

26. See Girard, *Deceit, Desire*, 55.

27. Of course, the prevailing view of infants may be changing, as indicated by the title of Alison Gopnik's recent book *The Philosophical Baby* (New York: Farrar, Strauss and Giroux, 2009).

28. See Hart et al., 60.

29. Girard, *Violence and the Sacred*, 18–19.

30. John Dollard, Leonard W. Doob, Neal E. Miller, O. H. Mowrer, and Robert R. Sears, *Frustration and Aggression* (New Haven: Yale University Press, 1939).

31. Amy Marcus-Newhall, William C. Pedersen, Mike Carlson, and Norman Miller, "Displaced Aggression Is Alive and Well: A Meta-Analytic Review," *Journal of Personality and Social Psychology* 78, no. 4 (2000): 670.

32. Marcus-Newhall et al., 671.

33. Girard, *Violence and the Sacred*, 9.

34. Marcus-Newhall et al., 682.

35. In reality, there are reasons other than fear that may cause someone not to lash out at the provoking agent. For example, frustration may build in a parent who refrains out of kindness from punishing a sick child for squalling.

36. See Marcus-Newhall et al., 670.

37. Simon Simonse, *Kings of Disaster: Dualism, Centralism and the Scapegoat King in Southeastern Sudan* (Leiden, Netherlands: Brill, 1992).

38. E. M. Beck and Stewart E. Tolnay, "The Killing Fields of the Deep South: The Market for Cotton and the Lynching of Blacks, 1882–1930," *American Sociological Review* 55, no. 4 (1990): 526–39.

39. Paul Dumouchel, "Mimétisme et genocides," in *Cahier René Girard*, ed. Mark R. Anspach (Paris: Éditions de L'Herne, 2008), 247–54.

40. Debra F. Horwitz, "Aggression between Dogs," Atlantic Coast Veterinary Conference, Atlantic City, New Jersey, October 9–11, 2001.

41. Cynthia M. Kahn, ed., *Merck Veterinary Manual*, 9th ed. (Rahway, NJ: Merck, 2005).

42. Peter C. Reynolds, *On the Evolution of Human Behavior: The Argument from Animals to Man* (Berkeley: University of California Press, 1981), 126–27.

43. Michael Tomasello and Malinda Carpenter, "Intention Reading and Imitative Learning," in *Perspectives on Imitation: From Neuroscience to Social Science*, vol. 2, *Imitation, Human Development, and Culture*, ed. Susan Hurley and Nick Chater (Cambridge: MIT Press, 2005), 133–48, (142–43).

44. Tomasello, in *Perspectives*, 146.

45. Characteristically, the Nazis were quick to revive it. In 1933, Hermann Goering instructed the Prussian police, "Shoot first and inquire afterwards, and if you make mistakes, I will protect you."

46. Marcus-Newhall et al., 673.

47. The memories are distorted by the fact that the original participants did not grasp the actual causal mechanisms at work, attributing all responsibility, both for the crisis and its resolution, to the scapegoat victim. Hence the distortions are not random; they follow predictable patterns that reflect the mistaken assumptions of the scapegoaters.

48. Girard, *Violence and the Sacred*, 97–98.

49. See Mark Rogin Anspach. *À charge de revanche: Figures élémentaires de la réciprocité* (Paris: Éditions du Seuil, 2002), ch. 1 (English translation in preparation).

50. Pierre Lemonnier, *Guerres et festins: Paix, échanges et compétition dans les Highlands de Nouvelle-Guinée* (Paris: Éditions de la Maison des Sciences de l'Homme, 1990), 99.

CHAPTER 8

Sacred Violence, Mimetic Rivalry, and War

Melvin Konner

To make these processes effective once again, people are tempted to multiply the innocent victims, to kill all the enemies of the nation or the class . . . and to sing the praises of murder and madness.

—René Girard

In brilliantly original works such as *Things Hidden since the Foundation of the World* (the source of the epigraph) and *Violence and the Sacred*, René Girard confronts fully a possibility that most modern social scientists have shied away from: that bloodshed may be at or close to the heart of all human social life.[1] The above quote occurs in the context of a conversation about the theories and movements spawned by Marx, Nietzsche, Freud, and even Foucault, all of which might be characterized as enthusiasms for which Girard has limited sympathy. Although they all share his willingness to acknowledge the role of violence, they also share the conviction that with the right approach (communist revolution, the triumph of the *übermensch*, universal psychoanalysis, or the overthrow of illegitimate power), the centrality of violence can be overcome.

I call this the Tinker Theory: Human life is terribly flawed, but if we tinker with the class structure, or the unconscious, or the reins of power, all will be well. Girard considers this naive and potentially dangerous. In reality, none of these approaches has succeeded in its goals, and in some cases the consequences have been unspeakably dreadful. Girard (rightly in my view) takes these failures as evidence that violence is and will likely remain central to human experience.

In fairness to Freud, some of his later writings—*Civilization and Its Discontents*, for example—seem almost Girardian in their acceptance of the ultimate tension between aggressive or "death" instincts and the cooperation needed for civilized life. But in his famous exchange of letters with Albert Einstein in 1932, it was Freud who played the optimist. The physicist began by bemoaning human susceptibility to propaganda leading to war: "How is it that these devices succeed so well in rousing men to such wild enthusiasm, even to sacrifice their lives? Only one answer is possible. Because man has within him a lust for hatred and destruction. In normal times this passion exists in a latent state, it emerges only in unusual circumstances, but it is a comparatively easy task to call it into play and raise it to the power of a collective psychosis."[2]

This is a great oversimplification, since the posited "lust for hatred and destruction" exists only under certain circumstances. A more general and easily evoked human emotional state is the anger that arises in response to frustration, fear, and grief. As we will see, combined with an easy slide into dichotomous thought that may lead to pseudo-speciation, the outcome can be ethnic violence, including war or genocide. Freud wrote of his "entire agreement" regarding the lust for destruction, but they differed on a crucial point: For Freud, "whatever fosters the growth of culture works at the same time against war."[3] Einstein doubted the civilizing power of culture, and to the world's great sorrow, he proved the more prescient thinker. Girard appears to be closer to Einstein, but for subtler reasons.

As I read him, he has made at least two major contributions to our discourse about violence. One is the concept of *mimetic rivalry*, according to which angry and competing individuals or groups in confrontation inevitably imitate one another, and in so doing escalate their rivalry into ever-greater risk of ever-greater violence. The other is the thesis of *sacrificial violence*, which holds that ritual sacrifice is a way of deflecting mimetic rivalry and exporting it from the community, defusing the process that otherwise results in what Hobbes called "the war of all against all." Whether impassioned and Dionysian, as in the *ad hoc* human sacrifice in Euripides' *The Bacchae*, or controlled by the strictest ritual, as in the priestly animal sacrifices in the Israelite Temple, the result is similar: the bloodshed is that of a designated victim, and it is sacred because it prevents us from shedding one another's.

And woe to the social world if it does not. Then, to paraphrase Mark Antony in *Julius Caesar*, you "let slip the dogs of war," and the foul deed of a sacrifice *not* agreed upon "cries above the earth with carrion men groaning for burial." Thus too in *Romeo and Juliet* do the Montagues and Capulets, "both alike in dignity," destroy each other piecemeal through interminable vendetta. Not even the accidental sacrifice of poor, good, funny Mercutio, the would-be peacemaker, deflects the violence; it goes on until the (also unintended) sacrifice of what each house loves most brings both down to indignity in a common plague of ultimate loss; "*all* are punishèd."

Much earlier in the history of tragic drama, the mimetic rivalry of Eteo-cles and Polyneices, two sons of Oedipus, annihilates his house as they tear Thebes apart and kill each other. Then, when the one is buried with honor and the other left to rot, Antigone too must die, sealed up in the earth, for the sisterly crime of burying Polyneices. And with her she brings down the whole house of the man who condemns her.

Surely, we think, if a ritual sacrifice could avert such endless mirroring of death breeding death, it would be a gift of the gods. But the role of sacri-fice is not always preventive; one-sided sacrifices can speed wars. Because in truth, "the face that launched a thousand ships and burnt the topless towers of Ilium"—and in the end toppled too the House of Atreus—was not that of Helen, but of Iphigenia, ritually slaughtered by her father Agamemnon, at Aulis, for the sake of wind.

Here we are closer to the dawn of civilization, but we are not there yet. Marx and Engels say in *The Communist Manifesto* that capitalism arose from the mud with blood oozing from every pore. This may or may not be metaphorically true of capitalism, but it is almost literally true of what we call civilization, which emerged from the mud of irrigated fertile land acquired and then protected by much slaughter.

Joining organized violence to religious zeal, early civilizations from the Yangtze to the Yucatan conquered and pacified large numbers of people who, through taxation and military service, provided resources for further expansion. Clashes with other, similar entities were frequent and inevitable. This dynamic has changed little in the thousands of years leading up to the modern age. We think we control the process, but human nature and human biology loom very large in the risk of ethnic violence and war.

In simpler ecological settings like that of the Nuer, a Nilotic people of southern Sudan, warlike tribal groups were able to form hierarchies of alliances and operate as organizations for predatory expansion. And in yet simpler and more static settings, people like the Dani and Enga of highland New Guinea and the Yanomamo of highland Venezuela sustain blood feuds and ritualized war over generations. Perhaps, as Girard suggests, the Kaingang of Brazil represented a degenerate form of this type of conflict, having slaughtered each other almost to extinction.

And yet it is possible to reach deeper, into the process of hominization, to find the origins of the violence at the heart of human life. Perhaps, in some ways, we transcended that background as we became human; perhaps ritual sacrifices helped us to do this. But in other ways we are all too similar to our pre-hominid ancestors for whom violence may have bred violence in an unending, bloody mimetic cycle.

ANIMAL VIOLENCE

Conflict is seen in all socially living animals, occurring over scarce resources such as food, space, or mates. But in the late twentieth century, our view of animal conflict changed. In the older view, threats and other aggressive displays were held to reduce actual violence by spacing individuals and arranging them in a stable hierarchy, and field research seemed to support this view.[4] Only humans, it was said, kill our own kind, because weapons distance us from our victims, making submissive displays and other natural restraints on violence ineffective.[5]

We now know this is false. It persisted in part because of lack of opportunity to observe animal killings. If a baboon troop had the same violent death rate as Americans do, it could take centuries to observe even a single killing.[6] But as field observations accumulated, deadly violence was seen in many species.

An example is competitive infanticide, first studied in Hanuman langurs.[7] At the core of langur groups are female relatives and their young; males may stay for a year or more but are ultimately transient. At some point new males appear, drive off the previous ones, and take their places. They soon kill all infants below six months of age; females resist for a time, but without success, and they soon become fertile again and mate with the new males. Various forms of competitive infanticide were described in chimpanzees, lions, wild dogs, and many other species.[8]

In this and other settings, violence evolved to help individuals and lineages sequester resources, including mates. Dominant males mate with ovulating females among baboons[9] and rhesus monkeys,[10] and competition for fertile females is a main cause of conflict. Male violence against females is also common in monkeys and apes, often in the context of sexual coercion.[11]

Chimpanzees show severe aggression, including attacks on females by the larger males, competitive infanticide by females, and violence between groups at territorial boundaries.[12] One or two victims temporarily separated from their own group are stalked and attacked by a group of males that beat, stamp, drag, and bite them to death. Females may be killed, but are more often absorbed into the other group. Entire groups have been eliminated by repeated, one-at-a-time ambush killings combined with female transfer.[13]

These chimpanzee ambush killings have now been studied in Uganda as well as in the Gombe Stream Reserve of Tanzania; the best predictor of such an attack is a critical number of adult males. In a group of 150 chimpanzees in the Kibale National Park studied for five years, the critical mass was about 18.[14] At about this number, the males would grow more excited until they went out into the forest in single file. On these expeditions they were uncharacteristically quiet, and they bypassed hunting opportunities along the way, until they crossed beyond the edge of their own territory. If they then came upon a lone male from the neighboring group, they collectively beat this victim; on five different occasions the beating was deadly.

But bonobos, which are as closely related to us as chimpanzees, do not show anything like such violence.[15] Genetic analysis will ultimately tell us the extent to which we share the violent genes of the one species, the nonviolent ones of the other, or a combination, and further field studies may suggest how much of bonobo nonviolence is situational; they are on the verge of extinction, perhaps in part due to their pacific natures. But at present, the best evidence for our greater similarity to chimpanzees is our own behavior.

FOSSIL VIOLENCE

For the early part of the protohuman fossil record, there is no evidence of violence, but there are only a few hundred specimens, mostly small parts of skeletons. The first hominids for whom there are many specimens are the Neanderthals, now viewed as probably off the line to modern humans, although they may have interbred with our direct ancestors. They are our closest relatives, and were behaviorally in many ways like us.

Neanderthal remains show a very high frequency of injuries, especially those found at Shanidar in Iraq, where the skeletons exhibit many healed and unhealed broken bones.[16] One man has a partially healed scar on the top of his left ninth rib due to a sharp object thrust into his chest.[17] He lived just a few weeks after the injury and may have suffered a collapsed lung. Another Neanderthal skeleton at the Skhul site has spear damage in the leg and pelvis.[18] Thus 40,000 to 50,000 years ago there is clear evidence of lethal violence in Neanderthals, and their high overall rate of injury may also be partly attributable to violence.

There is also clear proof of cannibalism.[19] In the cave of Moula-Guercy in Ardèche, France, Neanderthal bones dated to 100,000 years were butchered with the same techniques used on deer and goats. Other evidence suggests that the practice may be much older, and it has persisted up to recent times, sometimes associated with funerary ritual, sometimes with violence.[20] The later fossil record of modern humans prior to the invention of agriculture also shows scattered evidence of violence.

After that, there is no doubt: archeology has demolished "the myth of the peaceful savage."[21] In retrospect, we can see that the tenacity of this myth required substantial blindness to evidence, in accounts that were in effect "interpretive pacifications."[22] The record is sparse and therefore equivocal for pre-human species, but homicidal violence has clearly been part of our own species' way of life for at least 27,000 years.[23]

At Grimaldi in Italy, a projectile point was found embedded in a child's spinal column. At around the same time, Czechoslovakian cemeteries show considerable evidence of violent death, perhaps on a large scale. A Nile Valley man buried 20,000 years ago had stone projectile points in his abdominal section and another in his upper arm. There are many more such cases dating from between 14,000 and 12,000 years ago in Egyptian Nubia, and pre-agricultural sites in Europe show that violence was common, including the "Iceman" of 5,000 years ago, whose well-preserved body bears an arrow in the upper back.

All this violence took place during the hunter-gatherer stage of human prehistory, as much as 20,000 years before agriculture, and ethnography also shows that homicide occurs in many hunting and gathering societies, including the !Kung, Eskimo, Mbuti, and Hadza, among others.[24] It is often said that hunter-gatherers did not have group-level violence, but this claim is no longer sustainable. One cross-cultural study showed that almost two thirds of such societies had combat between communities at least every other year.[25] The sample in this study can be questioned—for example, it includes mounted hunters of the Great Plains, who are not suitable models for our past. Nevertheless, individual ethnographies of "classical" warm-climate hunter-gatherers show that their peacefulness has been exaggerated.[26] Other evidence includes southern African rock paintings, Australian aboriginal clubs and shields, and common spear wounds in 2,000-year-old skeletons in the American Southwest.[27]

With the spread of agriculture, archeological evidence of warfare in widely separate parts of the ancient world becomes decisive. Numerous skeletons show embedded arrow and spear points, left-sided skull fractures (caused by blows with weapons in the enemy's right hand), and parry fractures of the lower arm received while warding off such blows. Many sites include graves with weapons and armor; fortifications are found everywhere.[28] *The Iliad* and the biblical books of Judges and Kings reflect this continual clash of early civilizations. In fact, it is possible to view all history since the hunter-gatherer period as a process of ongoing, expansionist tribal warfare.[29] However, while settled agriculture may have intensified group violence, we probably had intergroup, and certainly had intragroup, conflict when we were still hunting and gathering.

From comparative ethnography, the emergence of true warfare appears to be associated with the transition from smaller to larger chiefdoms, which in turn led to the emergence of the state.[30] With growing populations, such features as social stratification, division of labor, and taxation become important. Alliances among religious, mercantile, and military hierarchies form the core of these societies, which continue to grow by conquest. As already noted, the latter process precedes the state; the Nuer became an effective organization for predatory expansion at the expense of their Dinka neighbors, despite the relatively modest level of social complexity in both rival groups.[31]

More centralized political systems such as those of the Aztec and ancient Mayans had greater military sophistication.[32] Such hierarchical societies increasingly resemble states rather than tribes or chiefdoms. This level of social organization corresponds to that of the legendary antagonists of the Bronze and Iron Ages, and from there to the antagonisms of modern states it is mainly a question of advancing technology.[33] Nationalism, as Arnold Toynbee put it, is new wine in the old bottles of tribalism.[34]

VIOLENCE IN SMALL-SCALE EXTANT SOCIETIES

Violent societies in the ethnological record strongly suggest that violence is simply part of human behavior.[35] Among the obvious examples are the Yanomamo of highland Venezuela, the Dani and Enga of highland New

Guinea, the equestrian Great Plains Indians of the United States, the Aztec, the Mongols, and the Zulu of nineteenth-century southern Africa. Such societies are found throughout the world, and they were demonstrably very violent. Twenty-five percent of adult male deaths among the Enga were due to violence, and their way of life was largely organized around it.[36] Among the Yanomamo, referred to as "the fierce people" by themselves and others, conditions were similar.[37] Forty percent of Yanomamo men had killed at least one other man, and those who had killed had more offspring than those who had not.[38] These and many other violent societies give the impression that we are a very bloody species composed of dysfunctional cultures.[39] Many older ethnographic accounts of warfare in primitive societies, including some that are now very pacific, suggest that ethnologists, like archeologists, have underestimated it.[40]

But consider also the least violent societies. Rates of homicide span three orders of magnitude among cultures, and such differences are important. The !Kung San of Botswana are often considered nonviolent,[41] yet their traditional homicide rate matched or exceeded that for American cities,[42] and there were also many assaults and fights between individuals.[43] While they have not engaged in group conflict in modern times, their contempt for other ethnic groups and even for !Kung in neighboring areas suggests that they have the psychological capacity for group conflict, and historical data show that they conducted violent intervillage raids in the past.[44]

Among the Semai, slash-and-burn gardeners of Malaysia, violence was said to be abhorrent and virtually nonexistent. "Since a census of the Semai was first taken in 1956, not one instance of murder, attempted murder, or maiming has come to the attention of either government or hospital authorities."[45] Childrearing and cultural ideology seemed to explain this:

> A person should never hit a child because, people say, "How would you feel if it died?" . . . Similarly, one adult should never hit another because, they say, "Suppose he hit you back?" . . . The Semai are not great warriors. As long as they have been known to the outside world, they have consistently fled rather than fight, or even than run the risk of fighting. They had never participated in a war or raid until the Communist insurgency of the early 1950's, when the British raised troops among the Semai, mainly in the west. . . . Many did not realize that soldiers kill people. When I suggested to one Semai recruit that killing was a soldier's job, he laughed at my ignorance and explained, "No, we don't kill people, brother, we just tend weeds and cut grass."[46]

But when the Semai became involved in British counterinsurgency against Communists in the 1950s, they showed a different side:

> Many people who knew the Semai insisted that such an unwarlike people could never make good soldiers . . . they were wrong. Communist terrorists had killed the kinsmen of some of the Semai counterinsurgency troops. Taken out of their nonviolent society and ordered to kill, they seem to have been swept up in a sort of insanity which they call "blood drunkenness." . . . "We killed, killed, killed. The Malays would stop and go through people's pockets and take their watches and money. We did not think of watches or money. We only thought of killing. Wah, truly we were drunk with blood." One man even told how he had drunk the blood of a man he had killed.[4]

After the war, Semai life returned to normal:

> Talking about these experiences, the Semai seem, not displeased that they were such good soldiers, but unable to account for their behavior. It is almost as if they had shut the experience in a separate compartment. . . . Back in Semai society they seem as gentle and afraid of violence as anyone else. To them their one burst of violence appears to be as remote as something that happened to someone else, in another country. The nonviolent image remains intact.[48]

This sequence of events could merely suggest what happens when men are bereft of their cultural context, and it could be precisely inexperience with violence that made their behavior extreme. Nevertheless, the Semai experience shows that upbringing and cultural ideology are only part of what determines the human tendency to violence.

Still, the differences among societies are important and may be explainable. In a cross-cultural study using the Human Relations Area Files and designed to sample representatively the ethnographic universe, matrilocal cultures had less warfare than patrilocal ones.[49] Conversely, where husband-wife intimacy is high—where husbands and wives eat together, sleep together, and share the child care—organized group conflicts are less common.[50] Cultures organized around frequent or intermittent warfare tend to segregate men, with separate men's houses for eating and sleeping, and often have men's societies in which boys may be initiated with substantial stress and actively trained for warfare. In general, we know that the social dynamic of male groups fosters violence,[51] and this appears to be an important process in recent terrorist actions.[52] Perhaps this can be thought of as a slower, more complex, language-based version of the dynamic described above for the building excitement of groups of male chimpanzees that have reached critical mass.

In concluding this section, it is very important to note a cross-cultural statistical study showing that after a society has been pacified by external powers, it becomes less interested in training boys to be aggressive.[53] This may not be surprising, but the empirical demonstration of it should serve to make us more cautious about calling cultures nonviolent on the basis of ethnographies done after, and sometimes long after, such pacification.

How Aggression Becomes War

I and others have reviewed the evidence for biological mechanisms of aggression elsewhere, and I will not repeat that here.[54] Suffice it to say that many kinds of both animal and human studies leave no doubt that physical violence has a strong genetic component, that it emerges in predictable ways developmentally with only partial dependence on the psychological environment, that it is instantiated in partly known brain circuits using partly known neurotransmitter dynamics, and that it is influenced by hormones, especially androgens, acting both during early development and in adulthood—a fact that accounts for the well-established and, in my view, clearly innate predominance of males in this kind of behavior. These biological factors, together with also proven psychosocial influences during development and in cultural context, will ultimately explain individual violence. But how does it become organized into group violence?

Three processes, each drawing on a strong human tendency that is separate from the tendency to violence, can be identified: (1) dichotomization or splitting of the social world, which Erik Erikson called "pseudo-speciation"; (2) emotional contagion and other processes of group psychology; and (3) following leaders. I will briefly consider these in turn, and the reader is referred elsewhere for more extended discussion.[55]

Splitting the Social World

As we have seen, and consistent with Girard's view, mimetic rivalries raised to the level of organized violent conflict have had a central role in human life. But they do not reflect only the proven human tendency to violence; they reflect another basic human trait: the tendency to dichotomize the social world. This is in part a special case of dualistic thought in general, identified by Mauss, Lévi-Strauss, and others as cross-culturally universal.[56] Dichotomies institutionalized in the language, religion, and customs of many cultures include night and day, human and animal, village and "bush," tame and wild, good and evil, male and female, right and left. The underlying perceptual reality may be a weak dichotomy or a continuum, but it is exaggerated or distorted by cognitive processes into an irreconcilable division.

This propensity is related to our low tolerance for ambiguity and cognitive dissonance.[57] In language, this is essential: there is a physical continuum between p and b, but we must know which one we are hearing in order to have meaning.[58] This may also apply to other aspects of cognition. As we evolved, it must often have been necessary to make decisions quickly, and this is made easier by having two clear choices. We must classify every stimulus as familiar or strange and decide on approach or avoidance. In the social realm, dichotomies of kin and non-kin, us and them, real people versus barbarians, heathen, Gentiles, or strangers are almost universal.

Such cognitive dichotomies also have an emotional dimension. Fear of the strange is a basic characteristic of complex nervous systems, and there is a continuum from attention through arousal to fear. Low-level electrical stimulation of the amygdala can produce alertness, while stronger stimulation in the same area will produce fear.[59] Novelty, depending on the degree and surrounding circumstances, produces either attention or fear in infants. In the social realm, the second half of the first year of life is a time of new distinctions, including attachment to a primary caregiver, and wariness or fear of strangers.[60] These reactions are ultimately the foundation of xenophobia in adulthood.

Classic studies in social psychology have traced the emergence and consequences of the us-them distinction. The Robbers Cave experiment brought 22 average eleven-year-old boys, all middle-class Protestants with similar educational backgrounds, to a summer camp, where they were randomly divided into two matched groups that differed in no measurable way. Despite joint activities and attempts to discourage competition, the groups began to compete, naming themselves, speaking disparagingly of each other, and reacting territorially to each other's "incursions." A tournament of formal competitions with trophies and prizes followed. Soon "good sportsmanship" gave way "to name-calling, hurling invectives, and derogation of the out-group to the point that the groups became more and more reluctant to have anything to do with one another." Gradually, "derogatory stereotypes and negative attitudes toward the out-group were crystallized." This process did prove to be reversible, but it is remarkable how quickly bigotry was created in two groups with essentially no differences between them.[61]

Similar findings have been repeated many times with adults and under a variety of more controlled conditions, confirming how easy it is to establish prejudice against arbitrarily formed out-groups, and to exacerbate the prejudice by giving the in-group members frustrating experiences or challenging their self-esteem.[62] Thus the social psychology of mimetic rivalry is well-grounded and fairly well understood.

Emotional Contagion

The fear and anxiety of even ordinary life in a complex and unpredictable world may be partly relieved by reducing responsibility for our actions. We do this by following rules, joining in collective action, or following a leader. Rules are probably the most benign of these tactics. Far more problematic is the mass or crowd psychology that sometimes arises in group action. According to Charles Mackay, in the nineteenth-century classic *Extraordinary Popular Delusions and the Madness of Crowds*.

> In reading the history of nations, we find that whole communities suddenly fix their minds upon one object, and go mad in its pursuit; that millions of people become simultaneously impressed with one delusion, and run after it, till their attention is caught by some new folly more captivating than the first. We see one nation suddenly seized, from its highest to its lowest members, with a fierce desire of military glory; another as suddenly becoming crazed upon a religious scruple; and neither of them recovering its senses until it has shed rivers of blood and sowed a harvest of groans and tears, to be reaped by its posterity. . . . Men, it has been well said, think in herds; it will be seen that they go mad in herds, while they only recover their senses slowly, and one by one.[63]

Mackay treats a wide variety of collective actions, including lynch mobs and witch hunts, reckless investment schemes such as the South Sea Bubble and the Tulip Mania, fads, pilgrimages, revolutions, and wars. Collective violence is seen in the context of susceptibility to emotional contagion, a phenomenon that has since been well demonstrated by psychologists.[64] We are not herd animals, but (by evolutionary history) members of small groups with complex social patterns. Yet even in our original small groups, the rudiments of these processes were present. Other classic studies in social psychology show that a person will deny the evidence of his or her senses, even with respect to something as simple as the relative length of printed lines, if a few others (stooges of the researcher) are in agreement against his judgment.[65] But beyond this, the hysterias Mackay describes may result in part from group sizes that violate the small-group dynamics we evolved with.

Group size notwithstanding, one frequent expression of mass psychology is the identification and destruction of enemies, which we may call *contagious enmity*, and which comes in two main forms. The first identifies weak internal enemies, then isolates and destroys them, as in the examples of lynch mobs, witch hunts, inquisitions, and genocide. The victims are called strange, confusing, evil, and dangerous to the spiritual and physical life of the larger group. In an extension of Girardian principles of sacrificial violence, killing them becomes a form of ritual purification.[66]

The second, perhaps more ominous form of contagious enmity identifies *external* enemies, similarly condemned but prepared for self-defense. If bloodshed is sacralized in primitive and ancient ritual, then the concept of holy war becomes more comprehensible; we send our sons and daughters into battle, and when they are killed, their blood makes the cause sacred—more so with each death. Sacrifices purify the community by exporting sins to the victim, but raiding and ambush killing of defenseless neighboring enemies may play an intermediate role. For example, Ilongot headhunting is directed against external enemies, yet "it involves the taking of a human life with a view toward cleansing the participants of the contaminating burdens of their own lives."[67] Through mimetic emotional contagion, the collective fear of two groups engaged in reciprocal contagious enmity will in the end be justified on both sides. That is, what may have been an irrational fear becomes a rational one as each side sees the real threat in the growing fear and hatred in the other.

Following Leaders

The infant's tendency to flee to a protective caregiver is bound up with fear, and if the infant's fear of strangers is transformed in adulthood into something like contempt, then the flight to a protector may take the form of obedience, conformity, chauvinism, or loyalty. A human group is not a herd and not merely a mob if it has a leader. Freud, in *Group Psychology and the Analysis of the Ego*—"group psychology" being a debatable translation of the German word *Massenpsychologie*—says that mass psychology operates in relation to a leader, yet something like mob psychology is still apparent: "the lack of independence and initiative in their members, the similarity in the reactions of all of them . . . the weakness of intellectual ability, the lack of emotional restraint, the inclination to exceed every limit in the expression of emotion and to work it off completely in the form of action."[68] But Freud does not restrict his model to popular delusions:

> We are reminded of how many of these phenomena of dependence are part of the normal constitution of human society, of how little originality and courage are to be found in it, of how much every individual is ruled by those attitudes of the group mind which exhibit themselves in such forms as racial characteristics, class prejudices, public opinion, etc.[69]

He views group psychology as not similar to, but "identical with" hypnosis, especially in "the behavior of the individual to the leader.[70] In his model, both the leader and fellow group members have a hypnotic power of suggestion. The flight to a protector—the "escape from freedom"—is to the dichotomous certitude of leader and group alike.[71] Freud's two main illustrations are armies and churches, both of which have an us-them distinction at their core; hypnosis occurs in relation to an enemy.

The submerging of individual will to authority is clearest in Stanley Milgram's famous experiments: Naive subjects were ordered to give presumed electric shocks to an unseen person. Most delivered what they believed were very dangerous shocks when ordered to by an authority figure in a white coat. "What is the limit of such obedience?" Milgram later asked. "We attempted to establish a boundary. Cries from the victim . . . were not good enough. The victim claimed heart trouble; subjects still shocked him on command. The victim pleaded to be let free, and his answers no longer registered on the signal box; subjects continued to shock him."[72] Encouragement by peers strengthened obedience.

Yet Milgram concludes that this is "not aggression, for there is no anger, vindictiveness, or hatred in those who shocked the victim. . . . Something far more dangerous is revealed: the capacity for man to abandon his humanity, indeed, the inevitability that he does so, as he merges his unique personality into larger institutional structures."[73] Philip Zimbardo's extremely disturbing experiments in which Stanford students role-played as guards and prisoners underscored the power of these processes, which made the one group as brutal as the other was cowed. As Milgram said, "This is a fatal flaw nature has designed into us, and which in the long run gives our species only a modest chance for survival."[74]

CONCLUSION

The concepts of mimetic rivalry and the sacredness of blood spilled in sacrifice, so clearly appropriate in literary and ethnological analysis, are also supported by scientific evidence that gives bloodshed a central role in human experience. Research on nonhuman societies, especially chimpanzees, suggests that a primordium of human ambush raiding, if not ritual sacrifice, is present there when a group of males gangs up on and kills a helpless victim. Many human cultures have violent raids with or without a ritual aspect, such as Ilongot headhunting and highland New Guinea raiding parties, in which multiple males ambush and crush a single victim. No doubt, in the process of hominization, language, religion, and ritual transformed what may have been blind killing by our chimp-like ancestors. Males predominate, emotional contagion operates, and both fellowship and leadership facilitate bloodshed. In humans, this can at times be directed at in-group members, but they cease to be that when they are split off from all that is human. Thus there may be a continuum from ambush raiding, to headhunting, to ritual sacrifice, to witch hunts and lynch mobs, and finally to genocide.

But if the detested enemy is not isolated and weak, but rather organized and strong, and you persist in attacking him, you have the peculiarly human outcomes of battle and war. Here mimetic rivalry emerges in full, as mutual fear and contempt are increasingly justified by events. Two mobs mirror one another in emotional contagion, and if they have leadership that compels or inspires obedience, you have a recipe for sustained or intermittent mutual slaughter, sometimes over generations. Emotional contagion can be a slow burn. Religious, ethnic, and national commitments justify the process on both sides, and with each death on one's own side—each sacrifice—the cause becomes more sacred. Aggressiveness is part of the story, but so are fear, contempt, dichotomization, emotional contagion, obedience, and the flight to the protector.

As I understand Girard, the reciprocal violence of equal combatants leads in the end to a sacrificial crisis. The mirroring brothers Polyneices and Eteocles slay each other, solving nothing, and then their armies must proceed with an endless mimetic slaughter. "Once violence has penetrated a community it engages in an orgy of self-propagation. There appears to be no way of bringing the reprisals to a halt before the community has been annihilated. If there are really such events as sacrificial crises, some sort of braking mechanism, an automatic control that goes into effect before everything is destroyed must be built into them. In the final stages of a sacrificial crisis the very viability of human society is put in question."[75]

And yet, as Girard notes, these matters are subject to much more severe and thorough repression today than are the sexual ones Freud dwelt on. Of the scores of thousands of modern behavioral and social scientists, only a few more than the handful cited here have addressed the problem of violence fully and deeply; yet the threat of it pervades our lives as a species, and we cannot learn to deal with it by drawing back. In the final lines of *Violence and the Sacred*, Girard foresees a modern sacrificial crisis:

> We have managed to extricate ourselves from the sacred somewhat more successfully than other societies have done, to the point of losing all memory of generative violence, but we are now about to rediscover it. The essential violence returns to us in a spectacular manner—not only in the form of a violent history but in the form of subversive knowledge. This crisis invites us, for the very first time . . . to expose to the light of reason the role played by violence in human society.[76]

NOTES

1. René Girard, *Things Hidden since the Foundations of the World*, trans. Stephen Bann and Michael Metteer (Stanford, CA: Stanford University Press, 1987); René Girard, *Violence and the Sacred*, trans. Patrick Gregory (Baltimore, Johns Hopkins University Press, 1977). The epigraph is from Girard, *Things Hidden*, 287.

2. Albert Einstein, "Why War? Letter to Sigmund Freud," in *Einstein on Peace*, ed. Otto Nathan and Heinz Nordan (London: Methuen, 1963), 190.

3. Sigmund Freud, "Why War? Letter to Albert Einstein," *Standard Edition of the Complete Psychological Works of Sigmund Freud*, vol. 22, ed. James Strachey (London: Hogarth, 1932), 287.

4. Vero Copner Wynne-Edwards, *Animal Dispersion in Relation to Social Behaviour* (New York: Hafner Publishing Co., 1962).

5. Konrad Lorenz, "What Aggression Is Good For," in *Animal Aggression: Selected Readings*, ed. C. H. Southwick (New York: Van Nostrand Reinhold, 1972).

6. Edward Osborne Wilson, *Sociobiology: The New Synthesis* (Cambridge, MA: Harvard University Press, 1975), 246–47.

7. Sarah Blaffer Hrdy, *The Langurs of Abu: Female and Male Strategies of Reproduction* (Cambridge, MA: Harvard University Press, 1977); Sarah Blaffer Hrdy, "Infanticide among Animals: A Review, Classification, and Examination of the Implications for the Reproductive Strategies of Females," *Ethology and Sociobiology* 1 (1979): 13–40.

8. Glenn Hausfater and Sarah Blaffer Hrdy, eds., *Infanticide: Comparative and Evolutionary Perspectives* (New York: Aldine-de Gruyter, 1984).

9. Irven DeVore, "Male Dominance and Mating Behavior in Baboons," in *Sexual Behavior*, ed. Frank Ambrose Beach (New York: John Wiley, 1965); Glenn Hausfater, *Dominance and Reproduction in Baboons (Papio Cynocephalus)* (Basel: S. Karger, 1975).

10. Kim Wallen and Pamela L. Tannenbaum, "Hormonal Modulation of Sexual Behavior and Affiliation in Rhesus Monkeys," *Annals of the New York Academy of Sciences* 807 (1997): 185–202.

11. Barbara Smuts, "Male Aggression against Women: An Evolutionary Perspective," *Human Nature* 3 (1992): 1–44; Barbara Smuts and Robert W. Smuts, "Male-Aggression and Sexual Coercion of Females in Nonhuman-Primates and Other Mammals: Evidence and Theoretical Implications," *Advances in the Study of Behavior* 22 (1993): 1–63.

12. Jane Goodall, "Infant Killing and Cannibalism in Free-Living Chimpanzees," *Folia Primatologica* 28 (1977): 259–82; Jane Goodall, *The Chimpanzees of Gombe: Patterns of Behavior* (Cambridge, MA: Harvard University Press, 1986); J v. L. Goodall, "Life and Death at Gombe," *National Geographic Magazine* 155 (1977): 592–621; Joseph Howard Manson and Richard W. Wrangham, "Intergroup Aggression in Chimpanzees and Humans," *Current Anthropology* 32, no. 4 (1991): 369–90; Richard W. Wrangham and Dale Peterson, *Demonic Males: Apes and the Origins of Human Violence* (Boston: Houghton Mifflin, 1996).

13. Jane Goodall, *The Chimpanzees of Gombe*.

14. Ann Gibbons, "Chimpanzee Gang Warfare," *Science* 304 (2004): 818–19.

15. Takayoshi Kano, *The Last Ape: Pygmy Chimpanzee Behavior and Ecology* (Stanford, CA: Stanford University Press, 1992); Richard W. Wrangham and Dale Peterson, *Demonic Males: Apes and the Origins of Human Violence* (Boston: Houghton Mifflin, 1996).

16. Eric Trinkhaus, "Hard Times among the Neanderthals," *Natural History* 87 (1978): 58–63; Eric Trinkhaus and William White Howells, "The Neanderthals," *Scientific American* 241, no. 6 (1979): 118–33.

17. Eric Trinkhaus, *The Shanidar Neandertals* (New York: Academic Press, 1995).

18. Steven LeBlanc and Katherine E. Register, *Constant Battles: The Myth of the Peaceful, Noble Savage* (New York: St. Martin's Press, 2003).

19. E. Culotta, "Neanderthals Were Cannibals, Bones Show," *Science* 286 (1999): 18–19; A. Defleur et al., "Neanderthal Cannibalism," *Science* 286 (1999): 128–31.

20. Peggy Reeves Sanday, *Divine Hunger: Cannibalism as a Cultural System* (Cambridge: Cambridge University Press, 1986); Paola Villa, Claude Bouville, Jean Courtin, Daniel Helmer, Eric Mahieu, Pat Shipman, Giorgio Belluomini, and Marili Branca, "Cannibalism in the Neolithic," *Science* 233 (1986): 431–37; Tim White, *Prehistoric Cannibalism at Mancos 5MTUMR-2346* (Princeton, NJ: Princeton University Press, 1992); David DeGusta, "Fijian Cannibalism: Evidence from Navatu," *American Journal of Physical Anthropology* 110 (1999): 215–41; Nicholas Wade, "If You Are What You Eat, Mind If I Move to Another Table?: Reconsidering Cannibalism," *New York Times*, January 2, 2000, section 4, page 3, New York City edition.

21. Lawrence H. Keeley, *War before Civilization: The Myth of the Peaceful Savage* (New York: Oxford University Press, 1996); LeBlanc and Register, *Constant Battles*.

22. Keeley, *War before Civilization*, 20.

23. Ibid., 37.

24. Richard B. Lee, *The !Kung San: Men, Women and Work in a Foraging Society* (Cambridge: Cambridge University Press, 1979); Bruce Knauft, "Reconsidering Violence in Simple Human Societies: Homicide among the Gebusi of New Guinea," *Current Anthropology* 28 (1987): 457–500.

25. Carol R. Ember, "Myths about Hunter-Gatherers," *Ethnology* 17, no. 4 (1978): 439–48.

26. Irenäus Eibl-Eibesfeldt, *The Biology of Peace and War: Men, Animals, and Aggression* (London: Thames and Hudson, 1979), 171–73.

27. LeBlanc and Register, *Constant Battles*, 100–127.

28. Keeley, *War before Civilization*; LeBlanc and Register, *Constant Battles*.

29. Andrew Bard Schmookler, *The Parable of the Tribes: The Problem of Power in Social Evolution* (Berkeley: University of California Press, 1983); John Keegan, *A History of Warfare* (New York: Vintage Books, 1993).

30. Timothy Earle, ed., *Chiefdoms: Power, Economy, and Ideology* (Cambridge: Cambridge University Press, 1991).

31. Marshall D. Sahlins, "The Segmentary Lineage: An Organization of Predatory Expansion," *American Anthropologist* 63 (1961): 322–45; Raymond C. Kelly, *The Nuer Conquest: The Structure and Development of an Expansionist System* (Ann Arbor: University of Michigan Press, 1985).

32. Keith F. Otterbein, *The Evolution of War: A Cross-Cultural Study* (New Haven, CT: HRAF Press, 1970).

33. Schmookler, *The Parable of the Tribes*; Michael Cook, *A Brief History of the Human Race* (London: Granta, 2003).

34. Arnold Toynbee, *A Study of History* (New York: Oxford University Press, 1972).

35. Paul Bohannan and American Museum of Natural History, *Law and Warfare: Studies in the Anthropology of Conflict* (Garden City, NY: Natural History Press, 1967); Keith F. Otterbein, *The Evolution of War: A Cross-Cultural Study* (New Haven, CT: HRAF Press, 1970).

36. Mervyn J. Meggitt, *Blood Is Their Argument: Warfare among the Mae Enga Tribesmen of the New Guinea Highlands* (Palo Alto, CA: Mayfield Publishing Co., 1977).

37. Napoleon A. Chagnon, *Yanomamo: The Fierce People* (New York: Holt, Rinehart and Winston, 1968); Napoleon A. Chagnon, *Yanomamö: The Last Days of Eden* (San Diego, Harcourt Brace and Co., 1992).

38. Napoleon A. Chagnon, "Life Histories, Blood Revenge, and Warfare in a Tribal Population," *Science* 239 (1988): 985–92.

39. Robert B. Edgerton, *Sick Societies: Challenging the Myth of Primitive Harmony* (New York: The Free Press, Maxwell Macmillan International, 1992).

40. Irenäus Eibl-Eibesfeldt, *The Biology of Peace and War: Men, Animals, and Aggression* (London: Thames and Hudson, 1979), 171–87.

41. Elizabeth Marshall Thomas, *The Harmless People* (New York: Vintage Books, 1959); Lorna Marshall, "Sharing, Talking, and Giving: Relief of Social Tensions among the !Kung," in *Kalahari Hunter-Gatherers: Studies of the !Kung San and Their Neighbors*, ed. Richard B. Lee and Irven DeVore (Cambridge, MA: Harvard University Press, 1976), 349–71.

42. Richard B. Lee, *The !Kung San: Men, Women and Work in a Foraging Society* (Cambridge: Cambridge University Press, 1979).

43. Marjorie Shostak, *Nisa: The Life and Words of a !Kung Woman* (Cambridge, MA: Harvard University Press, 1981); Marjorie Shostak, *Return to Nisa* (Cambridge, MA: Harvard University Press, 2000).

44. Irenäus Eibl-Eibesfeldt, *The Biology of Peace and War*, 171.

45. Robert K. Dentan, *The Semai: A Nonviolent People of Malaysia* (New York: Holt, Rinehart and Winston, 1968).

46. Ibid., 58.

47. Ibid., 58–59.

48. Ibid., 59.

49. Melvin Ember and Carol R. Ember, "The Conditions Favoring Matrilocal versus Patrilocal Residence," *American Anthropologist* 73 (1971): 571–94; William T. Divale, "Migration, External Warfare, and Matrilocal Residence," *Behavioral Science Research* 9 (1974): 75–133.

50. John W. M. Whiting and Beatrice B. Whiting, "Aloofness and Intimacy between Husbands and Wives," *Ethos* 3 (1975): 183–207.

51. Lionel Tiger, *Men in Groups* (New York: Random House, 1969).

52. Marc Sageman, *Leaderless Jihad: Terror Networks in the Twenty-First Century* (Philadelphia: University of Pennsylvania Press, 2008).

53. Melvin Ember and Carol R. Ember, "Prescriptions for Peace: Policy Implications of Cross-Cultural Research on War and Interpersonal Violence," *Cross-Cultural Research* 28, no. 4 (1994): 343–50.

54. Debra Niehoff, *The Biology of Violence: How Understanding the Brain, Behavior, and Environment Can Break the Vicious Cycle of Aggression* (New York: The Free Press, 1999); Melvin Konner, *The Tangled Wing: Biological Constraints on the Human Spirit*, rev. ed. (New York: Holt/Times Books, 2002); Melvin Konner, ed., *Human Nature, Ethnic Violence, and War* (Westport, CT: Praeger Security International, 2006).

55. Konner, *Human Nature, Ethnic Violence, and War*.

56. Mary Douglas, *Purity and Danger: An Analysis of Concepts of Pollution and Taboo* (New York: Praeger, 1966); Claude Lévi-Strauss, *The Savage Mind* (Chicago: University of Chicago Press, 1968); David Maybury-Lewis and Uri Almagor, *The Attraction of Opposites: Thought and Society in the Dualistic Mode* (Ann Arbor: University of Michigan Press, 1989).

57. Leon Festinger, *A Theory of Cognitive Dissonance* (Evanston, IL; Row, Peterson, 1957).

58. Roman Jakobson and Morris Halle, *Fundamentals of Language* (The Hague: Mouton & Co., 1971).

59. Holger Ursin and Birger R. Kaada, "Functional Localization with the Amygdaloid Complex in the Cat," *Electroencephalography and Clinical Neurology* 12 (1960): 1–20.

60. John Bowlby, *Attachment and Loss*, 3 vols. (London: Hogarth Press, 1969–1977); M. Lewis and L. Rosenblum, *The Origins of Fear* (New York: Wiley, 1973).

61. Muzafer Sherif et al., *Intergroup Conflict and Cooperation: The Robbers Cave Experiment* (Norman, OK: Institute of Group Relations, 1961).

62. H. Tajfel, *Social Identity and Intergroup Relations* (Cambridge: Cambridge University Press, 1982); Peter Robinson and Henri Tajfel, eds., *Social Groups and Identities: Developing the Legacy of Henri Tajfel* (London: Butterworth-Heinemann, 1997).

63. Charles Mackay, *Extraordinary Popular Delusions and the Madness of Crowds* (New York: Noonday Press, 1932).

64. Elaine Hatfield et al., *Emotional Contagion* (Cambridge: Cambridge University Press, 1994).

65. Solomon Eliot Asch, "Effects of Group Pressure upon the Modification and Distortion of Judgments," in *Groups, Leadership and Men*, ed. H. Guetzkow (Pittsburgh: Carnegie Press, 1951), 177–90.

66. See Girard, *Violence and the Sacred*; Robert Hamerton-Kelly, ed., *Violent Origins* (Stanford, CA: Stanford University Press, 1987).

67. Renato Rosaldo, *Ilongot Headhunting, 1883–1974: A Study in Society and History* (Stanford, CA: Stanford University Press, 1980), 140.

68. Sigmund Freud, *Group Psychology and the Analysis of the Ego* (London: The Hogarth Press, 1949), 81–82.

69. Ibid., 82.

70. Ibid., 78.

71. Eric Fromm, *Escape from Freedom* (New York: H. Holt, 1994).

72. Stanley Milgram, "Behavioral Study of Obedience," *Journal of Abnormal and Social Psychology* 67 (1963): 371–78; Stanley Milgram, *Obedience to Authority: An Experimental View* (London: Tavistock, 1974), 188.

73. Milgram, "Behavioral Study of Obedience," 371–78; Milgram, *Obedience to Authority*, 188.

74. Milgram, "Behavioral Study of Obedience," 371–78; Milgram, *Obedience to Authority*, 188.

75. René Girard, *Violence and the Sacred*, 67.

76. Ibid., 318.

CHAPTER 9

Desire, Mimesis, and the Phylogeny of Freedom

William B. Hurlbut

Thus there is no human nature . . . Man simply is. He is what he wills . . . One will never be able to explain one's action by reference to a given and specific nature—in other words, there is no determinism: man is free, man is freedom.

—Jean-Paul Sartre, *Existentialism and Humanism*

Who will deliver me from this body of death?

—Saint Paul, *Letter to the Romans*

Desire and its disordered dynamics is the central theme of René Girard's theory of mimetic process and its role in the foundations of human culture. Indeed, the early insights that led to mimetic theory were drawn from literary portraits revealing the unique character of human desire, its indeterminate nature, and its elaboration through imitation. These insights have been reaffirmed and extended through their application to anthropology, sociology, and economic theory, and, in turn, have greatly illuminated these fields of study. Now, with the new tools of social cognitive neuroscience, we may further expand our understanding of sociality, desire, and mimetic process, and their fundamental mechanisms in human neurophysiology and psychology.

By bridging and binding these disparate disciplines, mimetic theory may provide a crucial link in the quest toward a unified understanding of human life.

The Enigma of Desire

In explaining mimetic theory, Girard speaks of "that obscure thing named desire."[1] Indeed, no aspect of his theory is more opaque to the modern mind than the source of human desire and its ineluctable extension in mimetic rivalry—and the violence that follows. Girard at once acknowledges the natural phylogeny of desire as "grafted onto needs and appetites" evident in animal species, yet at the same time he affirms a decisive discontinuity in its human expression.

This ambiguity and obscurity is, indeed, reflected in the incongruity of prevailing academic (and popular) theories of human nature. From a naturalistic perspective, humankind is viewed as simply an extension and intensification of our animal origins, governed with an essentially mechanistic inevitability by the primary powers of pleasure and fear. On the other hand, from a moralistic perspective, we see ourselves as deracinated rationalities, untethered minds, like a creature so disembodied as to be capable of jumping over its own shadow.

Mimetic theory accepts neither of these perspectives. The crucial insight of mimetic theory is the connection it draws between desire, imagination, and the extraordinary human capacity for imitation and simulation. As Girard explains, "Desire is undoubtedly a distinctively human phenomenon that can only develop when a certain threshold of mimesis is transcended."[2] Acknowledging its foundations in natural patterns of need, he notes: "Human desire consists of the grafting of mimesis onto the instinctual patterns and the over-activation, aggravation, and disorganization of the latter."[3]

This connection between natural desire (need or appetite), mimesis, and the crisis of uncontainable conflict is the central assertion of mimetic theory and the missing piece of the puzzle in all other contemporary explications of human nature. Mimetic theory is not anti-natural. Rather, it expresses a more penetrating and persuasive explanation of what is everywhere evident but rarely recognized—that, unredeemed, the very freedom of humanity becomes its tragic fate.

Drawing on the theoretical and investigative tools of our advancing science, we can reconsider the origins of life, the foundations of freedom, and the dilemma of desire at the heart of human nature. Most specifically, we can seek a deeper understanding of the way freedom evolves inseparably from need (and desire), how imitation solidifies and extends the benefits of social life, and how the unique quality of human mimetic process transforms freedom and desire in a way that culminates in the crisis of conflict at the foundations of civilization. Clearly, as the tools of our advancing technology extend our powers for violence, such an inquiry becomes increasingly urgent.

The Phylogeny of Freedom

Potent but perishable, life in its very origins is precarious being. Unlike the inanimate elements from which they are formed, living beings are in active commerce with the world as the executive of their own existence. Desire, prefigured as need, forms the central axis of survival.

If we look back across the evolutionary panorama, we are at once struck by both its continuity and creativity—its stability and extension. Preservation within a changing world demands flexible adaptation, and adaptation brings forth new possibilities for the extension of the realm and reach of life. Within this crucible of creativity, one may discern a trajectory of ascent toward ever-greater powers of freedom, mind, and moral awareness.

At its most primary level, freedom within nature is prefigured as a widening range of response to the challenges and opportunities the environment imposes or provides. In the earliest organisms, vital powers of action and awareness encoded the most central biological values into reflexive responses in the service of life. Single-celled species, with a limited capacity to adjust, perdured through reproduction with mutation, but with multicellularity and the specialization of parts, an increasing capacity for flexibility and change provided a wider range of choice within a more distinct and bounded individual being.

With more complex forms, the primary vital powers of action and awareness came under the governance of desire, an inwardly felt sense of integrated and purposeful identity. The very definition of organism ("organ" in Greek means tool) implies the subordination and coordination of elemental parts into the cohesive unity of a larger, enlivened whole. Inwardly felt desire motivates, regulates, aligns, and sustains effortful engagement that bridges the span across time and distance to the object of action. Desire is embodied intention; it is the biological epicenter of being. As Leon Kass explains, "Appetite or desire, not DNA, is the deepest principle of life."[4]

With progressive evolutionary elaboration, ever more refined capacities for awareness and action, coordinated by desire, impelled and empowered the organism outward in active engagement with the wider world. Extension and multimodal coordination of the senses increased the clarity and precision of perception. Sight emerged as the dominant sense, delivering new tools of

comprehension and capacities of mind. Whereas smell required direct chemical contact, and sound gave formless information, sight allowed a knowing
and accurate encounter with the form and unity of wholes. It made possible
rapid perception of objects and actions at distant horizons, and an increased
understanding of the connections of action and reaction, of causes and effects.
This detached beholding of sight allowed a deeper and more accurate apprehension of the reality of things; sight allowed insight—fixed action patterns
gave way to genuine choice.

The new world relations made possible through the multifaceted perceptions of sight (form, color, texture, motion, cause, and connection) delivered a
dramatic reordering and extension of mind—and of the powers of prediction
and projection. The cerebral processing and storage of visual images allowed
detachability of object from image, making possible the mental manipulation and reordering of images—with the emergence of imagination and its
creative powers. In coordination with vocalization through the fine muscles
of the larynx, the capacity for imaging gave rise to symbolic representation,
genuine communication, and idealizing ideation. These powers, together with
the freed upper limbs and the "tool of tools," as Aristotle called the hands,
allowed a functional freedom and flexibility that has its psychic equivalents
in the open-ended desires and indomitable will of the human creature. The
omnivorous nature of our diet is paralleled by an equally omnivorous appetite
of dreams and desires.[5]

Social Being

These human capacities (and idealizing aspirations) are the culmination of
a trajectory of coordinated coevolution of ever more capable and conscious
animal kinds embedded in a unified ecological network grounded in active
communication. This coherent communion provides a dramatic extension of
the primary principle of freedom, allowing a deeper disclosure of the world
and a more articulate engagement with its multifarious forms. This in turn
provides the foundations for genuine sociality and the dramatic benefits that
ensue. A deeper exploration of the evolutionary foundations of social life can
provide insight into the inextricable link between desire, imitation, and personal identity—and the active aggression described by mimetic theory.

Among the earliest life forms, organisms drew information from one
another to pattern and coordinate basic biological functions such as reproduction and nurture. Gradually the direct chemical signaling suitable for aligning

collectives or swarms was transcended by more refined and specific mechanisms of individual communication. The externally evident demarcation of the head region, with its organs of sensory perception and communication, evolved in parallel with internal cerebral structures capable of processing more complex impressions of the environment and coordinating greater freedom of motion, including the active engagement of animal sociality.

With the ascent of mind, and a more distilled interiority of desire, came a parallel complexity of differentiation and integration of the external "look" of the animal. This "look," which is the literal translation of the Latin root of our word "species," is the product of a genetically determined plan as important as any internal vital organ. It provides the unity of form that reveals (or selectively conceals) the inner life of the organism.[6]

The progressive ascent toward more complex and integrated organization of the "inner life" and the external presentation of self reaches its fullest expression in the human form. Along with upright posture and the freeing of the hands (allowing an expanded range of communication) comes the reordering of the senses and the highly flexible, furless canvas of self-presentation we call the face. Upwards through mammalian evolution there is a progressive refinement of the structures of the face that facilitates increasingly subtle communication and penetration into the life of the other. With more than 30 finely tuned muscles of facial expression and vocal control, human beings are capable of a wide array of communicative expressions of emotions, intentions, and ideas.

These refinements of form and function establish the medium for the human capacity for empathy and genuine intersubjective sociality. This remarkable capacity in turn provides the foundation for language, continuity of culture, and the collective quest for the imagined ideal. To understand the social significance of empathy, and its role in mimetic process, it is helpful to explore the neurobiology and psychology of its basic mechanisms. This, in turn, will illuminate the inescapable connection between mimesis, desire, and human conflict.

Functional communication (and, of course, imitation) requires both a mode of effective transmission and a capacity for resonant reception capable of accurate interpretation. Without a shared platform of perspective, a common language of life, there can be no genuine communication, just stimulus and reaction. This essential measure of identity (however tenuous and limited at first) is made possible by the shared mechanism and meaning of our embodied being and our common experience within the world. Anton Chekhov is supposed to have said, "Everything I know about human nature

I learned from me." One might extend this with a small twist to say that "*Everyone* I know I learned from me." It is our shared perceptions, patterns of body motion, and inwardly felt emotions that provide the common link of meaning to the life of the other—and the toehold of truth and value beyond mere individual imagination.

As embodied beings, we are passionate and purposeful beings, whose intentions and values are grounded in our needs (and desires) and outwardly reflected in our dispositions, expressions, and actions. It appears there is an innate (and phylogenetically ancient) capacity for translating observed actions of another into the parallel postures and muscular motions of the self. This primary power of mimicry or copied behavior is common across the animal kingdom and has been observed in association with actions of central biological significance, including reproduction and foraging. Moreover, this observation and imitation involves more than patterns of motion; it includes the communication and assimilation of evaluation and its concomitant emotional expression.

To understand the innate connection between value and action, it is essential to consider the biological function of emotion. Emotions have their evolutionary origins in the physiological processes of body regulation. The postural and visceral changes in emotional states place the organism in a condition of readiness for action or response. The outward expressions of emotions reflect the self-preserving and purposeful disposition of the organism—they signal value and intention. The subjective feelings of emotions are evolution's later additions in the service of the inner life of consciousness and purposeful desire. This inseparable psychophysical unity of manifest emotion embodies the evolutionary experience of life's long history. Far from a private inner language of being, human emotions reflect survival strategies shaped by the physical and social parameters of our environment and shared with other members of our species, and indeed across life's larger process.

The universality of human experience and expression of emotions was called into question during most of the twentieth century, but has recently received support in the research of the psychologist Paul Ekman. Looking at more than a dozen cultures, including an isolated preliterate culture of New Guinea, he found a nearly universal language of facial expression of the emotions of anger, sadness, disgust, enjoyment, and surprise. In addition, he observed emotion-specific physiological changes in both the central nervous system (CNS) and the autonomic nervous system (ANS). And, most significantly for the role of emotion in communication, he noted that the very act of "voluntarily performing certain muscular actions generated involuntary changes in autonomic nervous system activity."[7] For example, accelerated heart rate and increased skin conductance accompanied the muscle actions expressive of anger. External bodily expression and inner meaning are inextricably linked.

It is this encompassing character (connecting action and affect) and shared quality of emotions between individuals that makes possible the process of empathy. It appears that even simulating the actions of another may establish an intersubjectivity that allows far deeper communication than mere observation and analysis. Studies of infant imitation and the discovery of mirror neurons may provide the psychological and neurobiological explanations of this extraordinary capability.[8]

Within 36 hours of birth, infants are able to discriminate some facial expressions and reflect them in the facial movements of their own brows, eyes, and mouth.[9] But how does this work? How does the infant know it is a mouth it is seeing, and how to move its own mouth in imitation? It appears that there is an innate ability to compare the sensory information of a visually perceived expression with the proprioceptive feedback of the movement involved in imitating the expression.[10] Taken together with the studies cited earlier showing that voluntary performance of muscular actions of emotional expression generated concurrent involuntary autonomic nervous system states, one can see the grounds for a genuine empathic resonance through facial communication and body posture.

These studies suggest an innate hard-wired connection between the sensory, motor, and visceral components of emotions. An emotional state such as anger, sadness, or surprise generates in an individual visible manifestations of body position and facial expression. Observing such postural and facial expressions subtly activates in the observer the same muscular movements and nervous-system responses that together constitute the physical grounding of an inwardly felt subjective state that would be represented by such expressions. We experience this, for example, when we see someone grimace in pain. Such a physical response in the observer translates into the corresponding emotion, thus establishing an empathically shared psychophysiological state between the observer and the one being observed.

The discovery of "mirror neurons" in monkeys (and suggestions from neuroimaging studies of similar cells in humans) may provide a physiological basis for such an explanation.[11] These cells fire not only when the individual makes certain motions (such as hand movements), but also when he observes others making the same motions. These cells, however, are not simply switches for mechanical actions. Indeed, they are more correctly described as convergence and divergence points in a broader circuitry of goal-oriented actions.[12] They are established within a comprehensive system that coordinates action, emotion, and intention—the essential components of functional communication.

Moreover, the auditory system also appears to be naturally primed for such affect-laden imitation. One-day-old infants exhibit inborn empathic distress reactions at the cry of other newborns. They respond with vocalizations that have the same auditory marks of genuine distress.[13] Further, researchers have found that, at least in adults, vocalizations of an emotional character can generate the concomitant emotion-specific autonomic nervous system changes. And here again (in monkeys), mirror neurons have been identified that respond to both production and perception of goal-specific sounds.[14]

These innate, hard-wired connections between the sensory, motor, and visceral components of emotions make possible a deep congruency of psychophysiological states between individuals. In earliest infancy, this binds the infant to the mother and provides a living link of natural communication and shared consciousness through which the infant develops the patterning for personal identity, and the platform for cultural awareness.[15] The philosopher Charles Taylor explains: "The genesis of the human mind is . . . not 'monological,' not something each accomplishes on his or her own, but dialogical."[16] All through infancy there is an interactive, empathic engagement between mother and child. Babies preferentially look at faces and follow the flow of the mother's emotional expressions and their vital association with the patterns of events. This ongoing reciprocity of interactions establishes congruency of emotional connection, nurturing the ties of attachment and the nonverbal foundations upon which language will later be built. At first the mother's spoken responses convey to the baby only feelings—the affective language of tone and prosody—but with continuing experience, this is extended to specific semantic content. With spoken language, there is functional participation in the broader cultural conversation that establishes meanings and values. Slowly the child is entrained to the society in which he is born—raised to the realm of a communally shared identity of beliefs, desires, ideals, and aspirations inaccessible to an isolated individual.

This unique human capacity of empathic intersubjectivity is the crucial threshold that Girard speaks of when he describes how mimesis transforms animal appetite and need into the full character of human desire.[17] Grounded in deep emotional resonance, mimesis conveys both evaluation and the urgency of action—lifting desire from a trigger of prepatterned need to an open-ended agenda inseparable from personal and social identity.

It is easy to recognize the biological benefits of such a system. Even in its earliest evolutionary expressions, imitation would serve an organism well. It bypasses the struggle of discovery and taps the experience of another; it is almost a form of parasitism, an economy where the rewards are reaped without the risks. Moreover, the special advantages for social existence are evident in the synergism and adaptive flexibility of coherent community. While individuals living alone may be able to exploit resources without competition, affiliation provides protective alliance, division of labor, and a longer period of childhood development. Empathic intersubjectivity binds the helpless infant to the mind of the mother and provides a direct line to the privileged information of inner states of desire and fear. Like a distilled pedagogy, it serves to entrain the developing child in the accumulated cultural values of its social group.

Yet, even as we acknowledge with a certain admiration (even awe) the intricate inventiveness of nature and the extraordinary possibilities that empathically grounded mimetic sociality provides, we cannot ignore the ruthless role of desire in the phenomenon of life. Sociality is as much a strategy of survival and self-promotion as are teeth and claws; shared values and common causes contain within their quest the seeds of conflict. The indeterminate desires and open-ended imagination of humanity is at once an invitation to freedom and the culminating crisis of creation. Discerning and defusing the destructive powers of mimetic desire is the most urgent task of our technological age.

The Neurobiology of Desire

To understand the dynamics of desire in its personal and social operations, we must complement the descriptions and observations of theology, philosophy, and the social sciences with the insights of neurobiology. New tools of investigation are clarifying the connection between desire and reward, and their role in human development and behavior. When we consider the significance of reward in learning, aligning intention, and motivating action—and the destructive power of its pathologies—it is surprising how little we understand about the basic biology of desire.

Early studies that seemed to identify a "pleasure center" have given way to a concept of distributed reward circuitry, and then to a wider integrated connection with purposeful intention and action. Clearly, the operation of desire in human life reflects the multidimensional complexity of personal identity—memory, imagination, and aspiration. Current terms of scientific inquiry parse the reward system into three cognitive and psychological components: associative conditioning in learning, affect and subjective hedonia, and motivational drive. Studies suggest several connections and conclusions relevant to reflection on desire and its role in mimetic process.

First, reward plays a primary role in learning by establishing and solidifying a system of anticipatory response and goal-directed action. Moreover, the associative connections that trigger response are pervasive but subtle, and the connection between learning and reward may be below the threshold of consciousness. In experimental settings, addicts will work for low dose levels of stimulants so small that they produce no subjective awareness and no measurable autonomic responses.[18] Thus reward, encoded as conscious desire or preconscious conditioning, plays a powerful role in shaping and promoting behavior, including behavior we cannot justify or explain. Moreover, it is now well established that the central circuits of reward and reinforcement also mediate preferences that are social in nature. And, as with the biology of addiction, this is true even when no reward is expected. Mimesis, in its disordered dynamics, may operate in a precognitive and prerational manner, below the threshold of mental intention or control.[19]

Second, there are neurobiological differences in the operation of two distinct dimensions of desire: "wanting" (the motivational state of active anticipation) and "liking" (the conscious awareness of subjective pleasure). Most specifically, the appetitive behaviors (wanting) differ from the consummatory behaviors (liking) in that they need not be accessible to conscious awareness.[20]

Liking appears to have a common neural mechanism for both natural pleasures such as eating and sex, and the direct action of pleasure-producing drugs. These neural structures include the orbitofrontal cortex, which mediates planning, evaluation, impulse control, and empathy. Likewise, cognitive "liking" involves the insula, a brain region crucial for body representation, regulation, and subjective emotional experience. The insula seems to mediate a mirror neuron–like connection between external and internal experiences essential for self-identity, admiration of others, and the shared intersubjectivity of the social emotions.[21] The insula is also involved in evaluations of disgust and responds to images of contamination and mutilation. This triangular connection between self, other, and images of mutilation may mediate the personal and social transformation from chaos to order effected by sacrifice of the sacred victim described by Girard in his theory of the scapegoat mechanism.

More fundamentally, before there are rewards, desires are alignments and intentions. Yet, motivations are more than the reasons and meanings empowering action; they are the very infrastructure of mind. Desire is essential to having a mental life at all.[22] The rare neurologic Athymhormic syndrome is characterized by an extreme passivity and loss of self-motivation, and is experienced as a complete mental blank unless stirred by external command. This suggests a fundamental role for desire as the grounding of personal identity—and the difficulty of dissociating its motivational imperative from the action (and aggression) of its realization.

The neurotransmitter dopamine, once described as the pleasure chemical, turns out to be neither necessary nor sufficient for the objective "liking" of natural taste observed in animals, nor the subjective pleasure reported with drug rewards in humans. Rather, dopamine's primary role appears to be in establishing "incentive salience," promoting a mere sensory input into a motivational magnet, and thereby solidifying the reward learning associated with "wanting" and the correlative states of attention and arousal that constitute motivation. These connections can be so compelling that monkeys will bite the pleasure lever that delivers reward.[23] The dominating and destructive power of addiction appears to be not so much a desired reward (liking) but a compulsive case of "wanting" driven by an overwhelming sense of defect and deficiency. Addicts generally hate their addiction, just as those caught up in mimetic rivalry seem compelled toward tragedy they neither choose nor understand.

One other aspect of the associative link between motivation and reward is relevant in connection to Girard's mimetic theory. While it is easy to see how preset biological needs (such as food and sex) and the natural pleasure of their consummation can feed back in a system of direct reinforcement, this is more problematic in mimesis, where no immediate experience of fulfillment is necessary (or, in some cases, even possible). It is now clear, however, that merely observing others experiencing states of arousal and reward will activate a resonant response; the desires of others quickly become our own. Even imagining reward (cognitive incentive representations) can do the same.[24] Clearly, the conceptual abstractions of desire, quite apart from their actual object, are adequate to incite the mirror of mimetic rivalry and its crisis of conflict.

The most extraordinary motivating power in human life is the idealizing imagination. Obviously this involves a neurological substrate for translating thought into action: without a connection between imagination and motivation there could be no truly creative action within the world. However fanciful or arbitrary it may be, the object of a culturally constructed ideal is a major organizing force in human society. Such ideals may be highly beneficial in constituting social solidarity and shared sense of purpose,[25] but they can very easily become the focus of pernicious interpersonal competition, corrosive to coherent community. It is immediately obvious that dopamine's dual role in both motivation and motor action provides a natural substrate for translating such competition to active aggression. Therefore, desire is not simply an abstraction, but an embodied power grounded in varied combinations of preconscious, mental, and emotional states expressed with a compelling drive in the direction of reward—and its consummation is inseparable from personal identity and its imagined ideals.

DESIRE AND MIMETIC CRISIS

One further insight from neurobiology may illuminate the connection between sociality, mimesis, and the crisis of mimetic rivalry described by Girard. While the indeterminacy in human desire and our idealizing imagination may imply that mimetic desire is untethered from biological imperatives, there is reason to think otherwise—at least if we reconceive the biological significance of sociality. The key to this enigma may be found in an evolutionarily ancient molecule, vasotocin (with a phylogeny of over 100 million years), and its more recent cousins oxytocin and vasopressin. These tiny neuropeptides play a powerful and pervasive role in a range of fundamental biological processes related to reproduction and sociality. In fish, during ovulation, vasotocin facilitates mating by reducing the female's natural fear of being approached by the male. In gregarious finches, a cognate peptide (mesotocin) influences affiliation and flock size. In voles, oxytocin and vasopressin play a crucial role in the intensity and permanence of pair-bonding, and in mice, maternal touching of neonatal offspring promotes the proliferation of oxytocin receptors, which later makes them more nurturing mothers with their own offspring. Likewise, in rats and monkeys oxytocin promotes nuzzling, licking, and reciprocal grooming.[26]

The basic mammalian reproductive and affiliative functions of oxytocin are further extended in human sociality—and suggest a biological influence on desires and preferences that appear from a common human perspective to be matters of purely personal choice or rationally grounded moral imperative. Oxytocin is released during sexual orgasm, maintains uterine contractions during childbirth, and plays a central physiological role in the production and release of breast milk.

These basic oxytocin-driven reproductive functions appear to set the relational foundations for broader dimensions of human sociality. Mothers who deliver vaginally (rather than by caesarian section) are, according to one brain-imaging study, significantly more responsive to the cry of their babies and appear to be less at risk of postpartum depression. "Vaginal delivery, but not caesarian section, involves the pulsatile release of oxytocin from the posterior pituitary which appears to affect brain regions that regulate emotions, motivations and postpartum mood."[27] Likewise, the earliest and most intense interpersonal experiences of the infant are in the context of lactation's oxytocin-driven bonding.

Throughout life, oxytocin appears to establish and sustain prosocial exchange. It facilitates cooperation, promotes trust, and reduces anxiety over interacting with strangers. Moreover, it establishes a whole new level of biological inclination and reward. Neuroeconomist Paul Zak explains, "Oxytocin constitutes a positive side of personal interactions; it literally feels good when someone seems to trust you, and this recognition motivates you to reciprocate." He goes on to explain: "Oxytocin causes the release of dopamine in deep midbrain regions associated with rewarding behaviors such as sex and food acquisition."[28]

This "pleasure" of social interaction appears to play a crucial role in binding children to parents, and later sustaining the social solidarity essential to cooperative community. While affiliation may feel from the inside to be a matter of choice, it is a compelling imperative of human existence, and the source of both our deepest pleasures and most profound moments of despair. Certainly there is no personal pain greater than that of stigma and social shame—the single most powerful force in reigniting the desperation of addiction.

This deep and abiding need, however, means that oxytocin is not simply the "cuddling hormone," as it is sometimes described, but a crucial element in establishing the medium of mimesis and its complex consequences in human life. Oxytocin potentiates social memory, including recollection of faces (but not inanimate objects); increases the intensity of social interactions; and strengthens the empathic intersubjectivity essential for true congruency of identity.[29] Primary patterns of biological need are increasingly subsumed and even displaced by more proximal and subjectively urgent social imperatives. Social approval and its correlative power become the dominant imperative of personal fulfillment—a kind of coinage for the realization and extension of self. The sense of social "connection" becomes more psychologically prominent than the content of that connection. Mimetic process colonizes and continuously reconstitutes the images and objects at the core of human desire, even to a degree at odds with basic biological needs. One might almost say that sociality is so prominent in human life that it transcends and displaces all other evolutionary imperatives—the "medium" (social connection) literally becomes the "message."

The pernicious influence of this process is everywhere evident in human life. The power of social pressure to entrain and enthrall the identity and life energy of the hapless individual is widely recognized. What is not fully appreciated is the more primary dynamic of mimetic rivalry and its destabilizing and destructive role at every level of human social life. Here again, oxytocin appears to play a role, but now in potentiating and intensifying conflict. Studies by the Israeli psychologist Simone Shamay-Tsoory show that intranasal administration of oxytocin increases envy and schadenfreude (gloating over other's misfortune) in a game of exchange—but only when the reciprocal player is perceived as arrogant and unconcerned with the welfare of others.[30]

Based on studies that show a positive mutual reinforcement between empathy, perspective taking, and perceived concordance of identity, it appears that a prominent component of sociality is a strong tendency for individuals to compare themselves with others.[31] When sociality is perceived to contravene the promise of reciprocal respect and the benefit it implies, disturbing dimensions of human nature are revealed—most forcefully in circumstances where the connection between admiration and identity of self seem at first to meld our very being into the unity of the relationship. Admiration promotes imitation, which draws desire into the mirror of mimetic identity formation, and, ultimately, into competitive rivalry. Disillusionment and disorder ensue.

This result correlates well with other human studies that suggest a role for the primary biological mechanisms of sociality in the root causes of aggression. Moreover, they seem consistent with observed differences in male and female sociality. In males, while oxytocin levels increase in response to female indications of trust, without it (with evidence of mistrust), levels of dihydrotestosterone spike, producing an aggressive response and promoting physical confrontation—a behavior not seen in human female subjects in equivalent circumstances of distrust.[32]

These bivalent responses seem consistent with an evolutionary explanation of sociality and its mimetic process—and with the natural connection between sociality, competition, combat, and mating. Mimetic desire may lead to a conflict over a mutually desired object of no particular biological value beyond its symbolic significance in personal pride and social status, but the reaction it provokes taps into an ancient evolutionary mechanism of violence in the service of self-promotion and self-defense. And, here again, the neurophysiology of reward may be a crucial component in the amplification of aggression. Animal studies have established that agonistic encounters are reinforced by dopamine-mediated mechanisms associated with reward properties—aggression feels good![33] Indeed, Shamay-Tsoory has observed that in rodents, oxytocin may be associated with an increased intensity of aggression.[34]

In nonhuman mammalian species, natural mechanisms of submission dampen the danger of escalating violence during confrontations over mates, territories, or scarce resources. In Syrian hamsters, agonistic behavior is directly modulated by vasopressin. An increase in vasopressin in the medial preoptic-anterior hypothalamus produces flank marking, a sign of territoriality and domination, but a decrease produces lordosis, the sign of submission.[35] Humans, however, do not exhibit such clearly preestablished programs of submission. Rather, as Girard has observed, at the foundations of human sociality where mimesis is overlaid onto the more primary evolutionary patterns of needs (and their social control), it provokes the "over-activation, aggravation, and disorganization of the latter"—leading to the uncontainable cycle of aggression, retribution, and reciprocal revenge.[36]

When the chaos of unconstrained violence threatens to endanger the entire community, primary psychological and social mechanisms emerge that solidify coalition and coordinate collective action. Atavistic impulses at the bedrock of the human psyche resurface, unmasking the guiltlessness of the predatory conscience. The foundational impulse of sociality is revealed, not as a rationally constructed social contract, but an instinctive cooperation centered on survival and rooted in the primary coordination (and unconstrained cruelty) of the hunt. The urgency of individual survival is riveted to social unity in a heightened arousal and intensity of collective mental and physical coordination that is at once wrought with terror and exhilaration.

The psychologist Victor Nell describes the Paleozoic roots of our primal proclivities for violence and its "savage joys," the fascination (even delight) associated with cruelty.[37] He argues that, rooted in the "emotional loading" of the evolutionarily ancient hunting adaptation, the primary psychological and physical mechanisms essential for successful hunting—high arousal and strong affect—establish what he terms the "pain-blood-death complex." He points to the dangers and high-energy demands associated with chasing and killing, and their associated pains of injury, stress, and extreme muscle fatigue. He suggests that powerful dopamine and endorphin-driven motivational and pleasure-producing mechanisms sustain and reward this behavior even in the earliest vertebrate predators. But while these physiological and psychological mechanisms of predatory adaptation served immediate instrumental ends for our animal ancestors, in the human species they are extended in a quest for personal and social power, evident in their elaboration in entertainment, social control, and sacrificial rites.

For the individual, the imperative of identity-defining desires links the very sense of personal survival to the competitive rivalry provoked by the mimetic mechanism. This is at once a crisis that isolates the individual and threatens community survival as the rivalry escalates to extremes. This crisis reaches toward its resolution only as the rivals are, by a seemly magical transformation, bound together in a powerful unity against a surrogate victim. The pain of isolation is dissolved within a ferocious frenzy of blood and death. The original object of mimetic rivalry is displaced by an exhilarating social solidarity potentiated by the arousal and heightened affect of the most primary predatory mechanisms of mind. Here the personal, social, and moral/spiritual are unified in the intensity of an act of righteous violence that carries the full symbolic power of blood and death—the predatory prerequisites of species survival. The frenzy of ferocity gives way first to horror, and then to the catharsis and calm of a reestablished order that seems to confirm the transcendent power and cosmological significance of the sacrificial rite. Life is restored by death.

CONCLUSION

From an evolutionary perspective, there are notable precedents for an initially beneficial adaptation such as mimesis to become amplified to a degree that it becomes dysfunctional and even threatens the survival of a species. Such patterns of "runaway evolution" have been proposed to explain extreme forms of mating display, such as peacock tails and giant antlers. Indeed, species specialization always carries the danger of overexploiting the immediate opportunity of a propitious adaptation at the expense of more general adaptability. It is ironic, however, that the very evolutionary strengths of the human species—our broad adaptability, functional freedom, and creative sociality grounded in mimesis—would culminate in a crisis of violence that threatens the entire order of creation. As our advancing technology delivers new and increasingly effective means of working our will in the world, we would be wise to more earnestly explore the biological basis of mimetic desire and the violence it promotes.

NOTES

1. René Girard, *Things Hidden since the Foundation of the World*, trans. Stephen Bann and Michael Metteer (Stanford, CA: Stanford University Press, 1987), 284.

2. Girard, *Things Hidden,* 283.

3. Girard, *Things Hidden,* 95.

4. Leon Kass, *The Hungry Soul: Eating and the Perfecting of Our Nature* (Chicago: University of Chicago Press, 1994), 48.

5. Kass, *Hungry Soul,* 70–74.

6. Leon Kass, *Toward a More Natural Science: Biology and Human Affairs* (New York: The Free Press, 1985), 325–30.

7. Paul Ekman, "Facial Expressions of Emotion: An Old Controversy and New Findings," *Philosophical Transactions: Biological Sciences* 335 (1992): 64.

8. Andrew Meltzoff and M. Moore, "Imitation of Facial and Manual Gestures by Human Neonates," *Science* 198 (1977): 75–78; Andrew N. Meltzoff, "The "Like-Me" Framework for Recognizing and Becoming an Intentional Agent," *Acta Psychologia* 124 (2007): 26–43; Andrew Meltzoff, "'Like me': A Foundation for Social Cognition," *Developmental Science* 10 (2007): 126–34; Vittorio Gallese et al., "Action Recognition in the Premotor Cortex," *Brain* 119 (1996): 593–609; Giacomo Rizzolatti et al., "Premotor Cortex and the Recognition of Motor Actions," *Cognitive Brain Research* 3 (1996): 131–41. See also Andrew N. Meltzoff, "Out of the Mouths of Babes: Imitation, Gaze, and Intentions in Infant Research—the 'Like Me' Framework," and Vittorio Gallese, "The Two Sides of Mimesis: Mimetic Theory, Embodied Simulation, and Social Identification," both in this volume.

9. Abraham Sagi and Martin L. Hoffman, "Empathic Distress in the Newborn," *Developmental Psychology* 12 (1976): 175–76.

10. Ibid.

11. Vittorio Gallese et al., "Action Recognition in the Premotor Cortex," *Brain* 119 (1996): 593–609.

12. Antonio Damasio and Kaspar Meyer, "Behind the Looking-glass," *Nature* 454 (2008): 167–68.

13. Sagi and Hoffman, "Empathic Distress," 175–76.

14. P. F. Ferrari, C. Maiolini, E. Addessi, L. Fogassi, and E. Visalberghi, "The Observation and Hearing of Eating Actions Activates Motor Programs Related to Eating in Macaque Monkeys," *Behavioral and Brain Research* 161 (2005): 95–101.

15. See Andrew N. Meltzoff, "Out of the Mouths of Babes: Imitation, Gaze, and Intentions in Infant Research—the 'Like Me' Framework," and Ann Cale Kruger, "Imitation, Communion, and Culture," both in this volume.

16. Charles Taylor, *The Ethics of Authenticity* (Cambridge, MA: Harvard University Press, 1991), 33.

17. Girard, *Things Hidden,* 283.

18. Kent Berridge and Terry Robinson, "Parsing Reward," *Trends in Neurosciences* 26 (2003): 507–13.

19. Timothy Behrens and Laurence Hunt, "The Computation of Social Behavior," *Science* 29 (2009): 1160–64.

20. Sagi and Hoffman, "Empathic Distress," 175–76.

21. B. Wicker et al., "Both of Us Disgusted in My Insula: The Common Neural Basis of Seeing and Feeling Disgust," *Neuron* 40 (2003): 655–64.

22. See Jean-Michel Oughourlian, "From Universal Mimesis to the Self Formed by Desire," this volume.

23. Behrens and Hunt, "The Computation of Social Behavior," 1160–64.

24. Ibid.

25. See Ann Cale Kruger, "Imitation, Communion, and Culture," this volume.

26. Paul Zak, "The Neurobiology of Trust," *Scientific American* 298, no. 6 (2008): 88–95.

27. James Swain et al., "Maternal Brain Response to Own Baby Cry Is Affected by Cesarean Section Delivery," *Journal of Child Psychology and Psychiatry* 49 (2008): 1042–52.

28. Zak, "The Neurobiology of Trust," 88–95.

29. Ulrike Rimmele et al., "Oxytocin Makes a Face in Memory Familiar," *Journal of Neuroscience* 29 (2009): 38–42.

30. Simone Shamay-Tsoory et al., "Intranasal Administration of Oxytocin Increases Envy and Schadenfreude (Gloating)," *Biological Psychiatry* 66 (2009): 864–70.

31. Claus Lamm, C. Daniel Batson, and Jean Decety, "Neural Substrate of Human Empathy: Effects of Perspective-taking and Cognitive Appraisal," *Journal of Cognitive Neuroscience* 19 (2007): 42–58.

32. Zak, "The Neurobiology of Trust," 88–95.

33. Maria Couppis and Craig Kennedy, "The Rewarding Effect of Aggression is Reduced by Nucleus Accumbens Dopamine Receptor Antagonism in Mice," *Psychopharmacology* 197 (2008): 449–56.

34. Shamay-Tsoory et al., "Intranasal Administration of Oxytocin," 864–70.

35. H. Elliot Albers et al., "Role of V1a Vasopressin Receptors in the Control of Aggression in Syrian Hamsters," *Brain Research* 1073–1074 (2006): 425–30.

36. Girard, *Things Hidden*, 95.

37. Victor Nell, "Cruelty's Rewards: The Gratification of Perpetrators and Spectators," *Behavioral and Brain Sciences* 29 (2006): 211–24.

CHAPTER 10

Naturalizing Mimetic Theory

Jean-Pierre Dupuy

MIMETIC THEORY AS SCIENCE

This chapter is about Mimetic Theory (MT) and its efforts to constitute itself as science. Its proponents know quite well that MT is *not only* a science. But if it is even partly a science, with as ambitious a goal as to account for everything from "the neuron to the eschaton,"[1] then it cannot shy away from confronting established scientific paradigms. Among its closest neighbors and potential rivals we find an emerging and powerful paradigm that results from the convergence of many disciplines: cognitive science, most especially cognitive psychology and cognitive anthropology; life sciences, in particular the neurophysiology of cognition; evolutionary theories and their many ramifications, in particular into the human sciences; the so-called "sciences of the artificial," most notably artificial intelligence and artificial life (aka synthetic biology); and a good chunk of the human and social sciences under the sway of the rationalistic paradigm (economics, game theory, rational choice theory, and the like)—not to mention the philosophical disciplines that cement those fields and hold them together, or the technologies that implement them, thereby increasing the power that human beings exert over the natural world, including themselves. For want of a better word, I will call this paradigm the "dominant paradigm," since its avowed ambition is to conquer the vast continent of the sciences of humanity *lato sensu*.

The encounter of MT with the discovery of "mirror neurons" illustrates the relative lack of preparedness of the former for this kind of confrontation. Many were the proponents of MT who enthusiastically and hastily claimed that it had found there its biological foundations and that its validity had thereby been proven. This claim is at best naive for at least two reasons.

In the first place, all but the most die-hard reductionists would agree that the proposition "Man is a super-mimetic animal" is self-evidently true even if the biological mechanisms responsible for it remain unknown. An analogy might be illuminating. After the works of such geniuses as Sadi Carnot, Lord Kelvin, and Clausius, the scientific world was certain that the so-called second law of thermodynamics (i.e., in an isolated system the state function known as entropy cannot decrease) was indeed a universal law of physics, and that, for instance, the efficiency of a steam engine producing work from heat was limited by an absolute threshold given by a certain formula. However, the attempts to account for this principle in terms of the properties of a collection of molecules whose degree of agitation could be measured by the Kelvin (or absolute) temperature— all those attempts that constitute what is known as statistical mechanics—failed miserably until fairly recently. But those were challenges for, and failures of, statistical mechanics, not at all for the second law of thermodynamics.

In 1933, Walter Benjamin wrote: "There is none of his [man's] higher functions in which his mimetic faculty does not play a decisive role."[2] The truth of this proposition is beyond any reasonable doubt. It is therefore a challenge for the neurosciences, and cognitive science more generally, *but not for MT*, to account for this fact. A would-be cognitive science that would prove incapable of it should simply be discarded.

Accordingly, in the last few decades, numerous researchers in the cognitive and social sciences have made significant strides in focusing on and illuminating the generative and foundational nature of human imitation or mimesis.[3] These more recent areas of research are significant for the necessary revisions their work requires of the "dominant paradigm" in cognitive psychology that was founded, more or less, without seriously taking the effects of mimetic processes into consideration. Can these more recent areas of research prove the validity of Mimetic theory, for instance via the discovery of neonatal imitation or mirror neurons? As I have just explained, I believe the problem to be poorly posed. To be sure, Mimetic theory might as well benefit from the discoveries of developmental psychology and neuroscience, but certainly not claim that the existence of say, mirror neurons proves its validity. We are dealing here with a situation that is commonly found in science, that is, the underdetermination of theories by facts.

I see the possible synergy between Mimetic theory and imitation research in cognitive science at another level, or rather in a radically different light. It is not a matter of one theory founding or grounding the other, but rather a matter of dialogue and listening to the multiple resonances that exist, *at a conceptual and formal level*, between the two fields, and trying to do something with those resonances and harmonies, compose a symphony, for instance.

Secondly, mainstream cognitive science did not wait for MT to become aware of the discovery of mirror neurons to appropriate it. Since MT is incompatible with at least some of the basic tenets of cognitive science, the challenge that it has to face is daunting. Rather than risking the ridicule of claiming possession of a good that unbeknownst to it someone else had already laid their hands on, MT must show that such early appropriation was made illegitimately. MT has no choice but to confront cognitive science and, more generally, the "dominant paradigm." Its academic future—which, to be sure, is not its sole concern—depends on this confrontation.

Mimetic Theory as Theory of Mind

Some Pitfalls in the Naturalization of the Mind

How the human mind emerged from the natural world is a fundamental issue that all theories of mind that purport to be scientific have to tackle. I will contrast two of them in this paper: analytic philosophy of mind inasmuch as it chose to ally itself with cognitive science; and the implicit theory of mind at the heart of MT.

In my philosophical history of cognitive science, I have defended the view that the naturalization and the physicalization of the mind attempted by cognitive science and analytic philosophy of mind is hampered by its very reliance on the folk-psychological account of human action in terms of the desire-belief model.[4] A theory of mind that sticks to this model is not amenable, I submitted, to natural laws. Therein reside the chances of MT to achieve what analytic philosophy of mind has not yet achieved, and probably will not, because it cannot. For MT is radically incompatible with the desire-belief model of human action, which it has fully demystified by showing that neither desires nor beliefs are autonomous.[5]

According to the desire-belief model, actions find their reasons *and* their causes in *mental states*, called "desires" (aka preferences) and "beliefs" (aka representations, etc.). The pertinent explanation of an action consists in showing that its description in the form of a proposition ("Peter grasps a bottle of beer," or in another version, "Peter decides to grasp a bottle of beer," or, still in another version, "Peter forms the intention to grasp a bottle of beer") is the conclusion of a practical syllogism, in which the major is a proposition that expresses a desire ("Peter wants to quench his thirst") and the minor is a proposition that expresses a belief ("Peter believes that drinking a beer will quench his thirst"). Ever since Bertrand Russell, such propositions expressing desires and beliefs are called "propositional attitudes."[6]

The problem encountered by the analytic or cognitivist paradigm is that it searches to "naturalize" the mind in a way that preserves something that ordinary psychology takes for granted; namely, that the contents of mental states have *causal* relevance in the explanation of our behavior. For a long time the obstacle seemed insurmountable, because philosophy of mind had convinced itself that the semantic content of a mental state, as described by its conditions of truth and of reference, depends on the entire physical and social environment of the subject; but if this content is supposed to have causal power in the physicist's sense, it can only be conceived in terms of the intrinsic properties of the mental state. It therefore appeared that the theory of mind could be naturalized only at the cost of depriving properties and mental states of all causal efficacy insofar as they are mental—thus making them pure "epiphenomena."

One of the most original theoretical proposals for overcoming this obstacle is the "anomalous monism" advocated by Donald Davidson in a famous 1970 article called "Mental Events."[7] Davidson postulated that every mental event is identical to a physical event; nonetheless, there is no identity relation between classes or types of mental events and classes or types of physical events. In holding that properties, whether mental or physical, are just such classes of singular events, this philosophical position combined an ontological monism (i.e., the view that ultimately there are only physical events) with a dualism of concepts and properties (i.e., that mental concepts are irreducible to physical concepts). Such a non-reductionist monism is "anomalous" in the following sense: while mental events cause other mental events as well as physical events, the relation of causality connects events with one another only insofar as they are events of the world, independently of whatever description, mental or physical, that we may give of them. The causal relation refers to a law of physics. By contrast, an explanation that involves mental concepts, after the fashion of the explanations furnished by ordinary psychology, cannot instantiate any strictly deterministic law in the physicist's sense—whence its non-nomological, or "anomalous," character.

In short, though mental or psychological concepts enjoy an explanatory autonomy, ontologically the relation of causality does not involve the mental insofar as it is mental. This remains an epiphenomenon (one hardly dares say "superstructure").

Reviewing the various attempts that have been made to give substance to a naturalistic and materialist theory of mind, notably the functionalism of a Jerry Fodor and anomalous monism, Pascal Engel comments:

> This dual concern with the reduction of mental concepts to concepts acceptable from the point of view of a scientific psychology and with support for the autonomy of such concepts well illustrates the permanent dilemma of a materialist theory of mind. Indeed, the more successful the reduction (which is to say, the more one manages to "explain" mental concepts in "physicalist" or "naturalist" terms), the less our usual mental concepts (those of common-sense psychology and of our pretheoretical conception of the mind) seem correct—and the more one is tempted to "eliminate" [mental concepts] in favor of [physical concepts], to hold that there simply are no such things as beliefs, desires, sensations, and so on. In other words, materialism ceaselessly oscillates between its "eliminativist" and "non-reductionist" versions. The project of a "naturalized" philosophy of mind similarly oscillates between these two tendencies.[8]

MT is resolutely non-reductionist, as we shall once again verify in the next section of this paper. To what extent it can afford being eliminativist, that is to say, non-mentalist, is still an open question. What can be said with certainty is that MT falsifies a basic assumption of the desire-belief model; namely, that beliefs and desires, if we decide to maintain this terminology, preexist the action. This could be shown in many ways. I will limit myself to two basic mimetic patterns: double mediation and pseudo-narcissism.

Double mediation obtains when the model imitates in the other the desire the other first found in him. This is all the more likely in a world in which there are few effective cultural barriers to rivalry, and in which each denies that he models himself on anyone else.

René Girard writes:

> In the world of internal mediation, the contagion is so widespread that everyone can become his neighbor's mediator without ever understanding the role he is playing. This person who is a mediator without realizing it may himself be incapable of spontaneous desire. Thus he will be tempted

to copy the copy of his own desire. What was for him in the beginning only a whim is now transformed into a violent passion. We all know that every desire redoubles when it is seen to be shared. Two identical but opposite triangles are thus superimposed on each other. Desire circulates between the two rivals more and more quickly, and with every cycle it increases in intensity.[9]

The mechanism is the following: A and B imitate each other reciprocally. A is anxious about B's desire, which alone can designate a target for his own desire. Some ephemeral and random sign makes him believe that B has designs on object O. Rushing to get there first, he thereby signals to his alter ego (B) the stakes of the rivalry. When B in turn imitates A's desire, the starting illusion becomes reality. The first to imagine the other's desire thus seems not to have been imagining at all: he now has the proof! This is a particularly interesting case of a self-fulfilling prophecy. Any object could have emerged from the mechanism. It all depends on how one enters into it. Now, the starting point of the process of the emergence of the object possesses an apparently contradictory twofold property: it is nothing or almost nothing, a *je ne sais quoi*, a caprice, a chance occurrence; and yet, it plays a crucial role, since everything takes place as if it were the thing that "determined" the object, the "objective" reality that is to emerge. There is a beginning, but that beginning is evanescent; there is determinism, but the determining factor is in the final instance ... beyond grasp.

Mimetic rivalry and the object of conflicting desires determine each other; neither preexists the other. The actions that reveal the converging desires cause them to emerge and intensify. Desire does not preexist action. There is a feedback loop of action onto desire.

To introduce pseudo-narcissism, there is no better guide than the famous passage in which Girard scathingly and humorously critiques Freud's theory of narcissism:

> The coquette knows a lot more about desire than Freud does. She knows very well that desire attracts desire. So, in order [to] be desired, one must convince others that one desires oneself. ... If the narcissistic woman excites desire, this is because, when she pretends to desire herself and suggests to Freud a kind of circular desire that never gets outside itself, she offers an irresistible temptation to the mimetic desire of others. Freud misinterprets as an objective description the trap into which he has fallen. What he calls the self-sufficiency of the coquette, her blessed psychological state and her

impregnable libidinal position, is in effect the metaphysical transformation of the condition of the model and rival. ...

The coquette seeks to be desired because she needs masculine desires, directed at her, to feed her coquetry and enable her to play her role as a coquette. She has no more self-sufficiency than the man who desires her, but the success of her strategy allows her to keep up the appearance of it, since it offers her a form of desire she can copy. ... To sum up: in just the same way as the admirer caught up in the trap of coquetry imitates the desire that he really believes to be narcissistic, so the flame of coquetry can only burn on the combustible material provided by the desires of others.[10]

We are dealing here with a variant of double mediation: the coquette's desire for herself is mediated by those she attracts, while their desire for her is mediated by what they believe is her purely independent self-desire. Here, it is the *belief* that the other is self-sufficient that is generated by what it causes, namely, the revelation through some action that one desires her. Belief does not preexist action. There is a feedback loop of action onto beliefs.

Those descriptions still preserve the mentalist terminology of folk psychology. MT has not yet crossed the line, which would consist in jettisoning it altogether. Its roots in the analysis of literature probably stop it from doing so. Is it prepared to take that step, if that is the price to pay in order to become amenable to naturalization? What follows can help think through this crucial issue.

Away from Mentalism, Closer to Mechanism

What is remarkable is that within cognitive science itself, a trend has always been present that tends to deconstruct what a Heideggerian would call the "metaphysics of subjectivity." Cybernetics played an essential role here by showing that the fastest route to naturalizing the mind was to mechanize it.[11]

I will take a detour through the social sciences to illustrate this point. Methodological individualism in the social sciences has one golden rule: never treat aggregates as subjects. To do so would be to commit a category error. It is an error that we are tempted to commit constantly. Consider voting in elections, which today is the democratic procedure par excellence. Whether it is a matter of electing a president through universal suffrage, or of a referendum requiring the voter to answer yes or no to a certain question, it frequently happens that the distribution of votes is nearly equal: the winner prevails by very little. This state of affairs occurs so frequently, in fact, that a vote massively in favor of one of the options proposed inevitably arouses the suspicion that there was something irregular about the voting procedure. Since the time of Montesquieu, democratic theory has multiplied the number of attempts to account for this observation. For an information theorist, however, an even vote is one that maximizes entropy, which is to say disorder. In formal terms, the procedure is equivalent to a lottery.

This interpretation may seem surprising, but it acquires still more force if one is aware of what rational-choice theorists call the "voting paradox." Except in the extremely improbable case where ballots are equally divided between two options, the conclusion is unavoidable from an individualist perspective that no single ballot cast by any of the electors has any effect whatsoever: the effect of any individual vote upon the outcome is, strictly speaking, nil. The individual voter is forced to accept that the answer to the question "Would the final result have been altered had I voted otherwise than I did?" is no. But what do we observe in practice? We observe that the outcome of a vote is often interpreted as the manifestation of the carefully considered choice of a collective subject: the "people," the "electorate," and so on. In the case of a referendum that goes in favor of the "yes" by an extremely narrow margin, the usual interpretation is something like the following: "The electorate, in its wisdom, has answered yes to the question put before it, but it has also wished to issue a warning to all those who seek to move too fast." It is as though one had made a subject out of chance—a collective subject in what might be called a position of exteriority to itself, since obviously the electorate is supposed to transcend each citizen taken individually. In the life of the innumerable committees and commissions to which modern societies confide responsibility for administering public affairs, the resort to anonymous voting is very frequently only a disguised means of delegating to chance a decision that presumptively rational debate has shown itself incapable of reaching. But these forms of randomly generated consensus are considered legitimate and meaningful to exactly the extent that they produce an external or transcendent point of view, and so can be taken as the decisions of a collective subject. Those are cases of generation of *transcendence from immanence*.

This interpretive attitude—what cognitive philosopher Daniel Dennett calls the "intentional stance"—is an inevitable fact of the human condition.[12] We ceaselessly attribute to others "mental states" (intentions, desires, beliefs, and so on), no matter whether the other is a human being, an animal, a machine—or a human collectivity. As a practical matter, this stance tends to weaken methodological individualism (or, as I would rather say, to make it more complex) since the individual subject no longer has a monopoly upon certain attributes of subjectivity. It becomes necessary to admit, in addition to the existence of these individual subjects, the existence of quasi-subjects, which is to say, collective entities capable of exhibiting at least some of the attributes that one had thought were restricted to "real" subjects—individuals—and in particular, the existence of mental states. One thus does not hesitate to say of an organization, or more generally of a collective entity, not only that it is capable of learning, but also that it is capable of knowing, remembering, analyzing a situation, making experiments, forming concepts, making decisions, and acting.

Now, a whole tradition in cognitive science, from the time of its cybernetic origins up through the present day, has presented a picture of the individual subject itself as a quasi-subject, that is, as a collective entity manifesting the properties of subjectivity. When I think, remember, desire, believe, decide, and so on, the subject of these predicates is not a ghost in the cerebral machine—a concealed homunculus, as it were—it is the machine itself, in the form, for example, of a network of neurons. According to this point of view, there is, as Dennett insists, no ghost in the machine, no center of subjectivity. The attributes of subjectivity are emergent effects produced by the spontaneous, self-organized functioning of a complex organization in the form of a network. Cognitive scientists who defend this thesis, or a variant of it, resort sometimes to curious expressions: Francisco Varela speaks of "selfless selves,"[13] Daniel Dennett of non-selfy selves,[14] Marvin Minsky of a "society of mind."[15] But the idea is clear enough.

All of this is to say that the weakening, indeed the "deconstruction" of the metaphysical (i.e., Cartesian and Leibnizian) concept of subjectivity took place at the intersection of the social and cognitive sciences on both a "macro" and a "micro" level. On the "macro" level, the attributes of subjectivity are not the monopoly of individual subjects: collective entities can exhibit them as well. On the "micro" level, the attributes of subjectivity are not attributes of individual subjects: they are emergent effects produced by the functioning of *subject-less* processes. In both cases, the tool used to deconstruct the subject is the same: the realization that a complex network of interactions among simple entities—formal neurons in the case of the individual quasi-subject, schematic individuals in the case of the collective subject—can exhibit remarkable properties. For someone like Dennett, it is neither more nor less justified to attribute a mental state, such as an intention, to a human being than to a collectivity.

If one reads contemporary cognitive philosophy on consciousness and the self, one cannot help but be struck by its constant references to social and political metaphors, even electoral ones. The emergence of a self at a given moment, in the course of one of those processes that constitutes the life of the mind, is likened to the transition from being in a crowd—a crowd of mental events, of neuronal configurations—to being in an organized political community, created by the election not of a center of control but of a representative: the "head of mind," as the individual "subject" might be called, playing the same role as the head of state. In this scenario, mimetic crises occasionally arise: several potential representatives emerge as rivals for power, and so on. *Mental mechanisms and social mechanisms are placed on the same level, precisely to the extent that they are mechanisms.*

Therefore, social theory and cognitive science, in their recent developments, converge, or rather grope towards, two fundamental conclusions. Firstly, the autonomy of the human subject posited by classical modern philosophy, from Descartes to Leibniz, is just an illusion ("Descartes' error," to quote from Antonio Damasio.)[16] Secondly, no center of control is responsible for the transition from disorder to order in the case of human collectivities.

Those findings receive an incredibly strong and novel light from MT. The human subject is radically incomplete, insufficient, for being radically mimetic. The scapegoating mechanism is responsible for the transition from violent disorder to a form of order that is always fragile and bound to collapse; and the victim is surely not in a position to control that process. The important word here, a talisman that can help bring MT and parts of the dominant paradigm closer to one another and to a position from which they can engage in a fruitful dialogue, is "mechanism." We are now going to confirm this intuition by confronting MT with Evolutionary Theory (ET).

MIMESIS AS MORPHOGENETIC PRINCIPLE

Mimetic Theory and Evolutionary Theory

MT includes a theory of hominization and of cultural evolution, which presupposes a certain theory of biological evolution. The latter is yet to be conceptualized, another daunting task for MT. Most neo-Darwinian theories, which set themselves the task of accounting for the passage from animal to man, characterize the latter in mentalist terms. This is not what MT as evolutionary theory is interested in doing, for the reasons just adduced. How, then, can the conversation between MT and ET be engaged?

ET plays an essential, structuring role in what we have called the dominant paradigm. It itself has had to avoid many pitfalls in its conquering enterprise, the most glaring one being what goes by the name of "Social Darwinism"—that is, the careless application of "Darwin's dangerous ideas," such as "struggle for life" or "survival of the fittest," to the social domain. The thinker who has best understood the absurdity of neo-Darwinian evolutionary anthropologists or psychologists brutally applying to social phenomena what they think is the case in the biological realm, was the social philosopher and Nobel Prize laureate in economics Friedrich Hayek. In a text that has become a classic of the epistemological literature on evolutionary thinking, he criticized:

> the erroneous belief that it is a conception which the social sciences have borrowed from biology. It was in fact the other way round, and if Charles Darwin was able successfully to apply to biology a concept which he had largely learned from the social sciences, this does not make it less important in the field in which it originated. It was in the discussion of such social formations as language and morals, law and money, that in the eighteenth century *the twin conceptions of evolution and the spontaneous formation of an order* were at last clearly formulated, and provided the intellectual tools which Darwin and his contemporaries were able to apply to biological evolution. Those eighteenth-century moral philosophers and the historical schools of law and language might well be described ... as *Darwinians before Darwin*.[17]

Hayek cruelly added, "A nineteenth-century social theorist who needed Darwin to teach him the idea of evolution was not worth his salt. Unfortunately some did, and produced views which under the name of 'Social Darwinism' have since been responsible for the distrust with which the concept of evolution has been regarded by social scientists."[18]

Among the Darwinians before Darwin whose thought had a powerful influence on the author of *On the Origin of Species*, Hayek singled out the Scottish Enlightenment in general and Adam Smith in particular. It was not the economist or the fledging discipline called political economy that had the major impact on Darwin's thinking. The book that interested him most was the treatise in moral philosophy that Adam Smith published in 1759—that is, exactly one hundred years before *On the Origin of Species*. This book was titled *The Theory of Moral Sentiments* (TMS), and it preceded by 17 years the publication by the same author of the first work of modern economics, *An Inquiry into the Nature and Causes of the Wealth of Nations* (WN). It must be added that Smith always considered the former, TMS, a much superior work, where the key to the latter could be found.

What did Darwin find so essential in TMS? He found something absolutely stunning: a remarkable discovery equal to the most fabulous accomplishments of the human mind. He found that it was possible to conceive of a complex order, of its genesis and evolution, without any recourse to the postulation of a designer, God or Man. Another member of the Scottish Enlightenment, Adam Ferguson, had used a memorable formula: "Social order is the result of human action but not of human design." Darwin saw there how he could conceive of a natural order, capable of complexifying itself ever more, without having to posit that a demiurge, a grand clockmaker, or a grand architect planned or designed it according to His will. That discovery was overwhelming and had nothing to do with the economic frivolities that go today by the name of Social Darwinism and are the result of the brutal application of biological ideas to society: the market is the place where the weakest are eliminated and only the fittest survive, capitalism is synonymous with struggle for life, and the like.

Smith's account was a brilliant precursor of what would be called, in the second half of the twentieth century, theories of complex, self-organizing systems, autopoietic systems, or, in Hayek's terminology, spontaneous orders. It turns out that MT itself is a theory of self-organizing complex systems, no less than Darwinism and ET. The mechanisms that it analyzes are *morphogenetic*: they are capable of generating new forms. They are simple, but their simplicity brings about complexity. My methodological advice is then the following: Let us set up the dialogue between MT and ET at the level of the *formal models* that structure the one and the other. We will avoid the many pitfalls that await those who carelessly smuggle biological notions into the social and cultural realms and vice versa, and we will focus on the interesting questions: it is likely that biological self-organization as seen by ET, and social and cultural self-organization as seen by MT share fundamental traits and differ in their material implementations. Let us explore systematically what they are. In what follows, I will present a brief illustration of what this research program might look like.

Chance and Necessity: Two Principles of Evolution

Both MT and ET have isolated *morphogenetic* principles of organization, which are neither biological nor economic or social per se, but reside at a higher level of abstraction.

I will start from MT. I would like to explore the reason that makes mimesis a morphogenetic principle of such power as to generate self-organizing systems whose evolution is endogenously determined. We know that evolution is a mix of chance and necessity. This characterization is too general, though, and a fundamental distinction must be drawn between two principles of evolution: the first one is dubbed "order from noise" and the second "complexity from noise."[19] To illustrate their difference at a formal level, I will present two very simple mathematical thought experiments.

The first experiment has been in operation at the Palais de la Découverte, Paris's science museum, ever since its foundation in 1937. All visitors are invited to participate. They are requested to cast a needle onto a grid of equidistant lines. The length of the needle is half the distance between two neighboring lines. Either the needle intersects one of the lines or it does not. The setting is electrified, which permits a counter to compute the frequency of the cases in which there is an intersection. Over time, tens of millions of visitors have cast their needle, and the proportion of intersection cases has oscillated around and converged towards a value that is now determined with thousands of decimals: the beginning is 0.318309886183791 ... It turns out that this value is the inverse of pi (pi being the ratio of any circle's circumference to its diameter). It is thus that the value of pi can be *experimentally* determined with a precision that is, if we wait a sufficiently long time, as high as one wishes. The same experiment is being reproduced in many other science museums of the planet, for instance at the San Francisco Exploratorium. Everywhere the process converges towards the same value.

This experiment, known as "Buffon's needle," as spectacular as it may seem, is just an illustration of the so-called "law of large numbers," one of the pillars of the probability calculus: the frequency of a random event tends over time towards its a priori probability. The French naturalist Buffon, who was also an eminent mathematician, could demonstrate in a highly elegant way that the a priori probability of an intersection in the experiment at hand is the inverse of pi.[20] Chance, here, is the instrument through which a preexisting necessity realizes itself. This is a case of "order from noise."

The second thought experiment illustrates the incredible morphogenetic power of imitation. Dubbed the "Polya's urn scheme," after the name of Hungarian-born Stanford professor of mathematics George Polya, it has become the matrix of a wide variety of scientific models.[21] An urn contains one white ball and one black ball. One ball is drawn randomly from the urn and its color observed; it is then placed back in the urn together with another ball of the same color. Hence the number of balls in the urn increases by one every time. The question is, how does the proportion of white balls, say, evolve over time? It is very easy to simulate this evolution with a simple calculator coupled with a generator of random numbers. We realize the experiment and observe with surprise that the system seems to have the same kind of dynamics as the Buffon's needle case: a series of oscillations flattens out and converges towards a certain value. A second surprise is that this value is not 0.5 (in which case half the balls in the urn would be white). Why is it a surprise? Because the setting is perfectly symmetrical. The observed *breach of symmetry* seems to come from nowhere. No rational explanation seems capable of accounting for it.

Note what makes this model the simplest formalization of a mimetic dynamic. Every random event (here, the draw of a ball of a certain color) changes the conditions for the next draw by reinforcing the odds of the color in question. This self-reinforcing process is very much akin to the mimetic pattern that Girard calls "double mediation": imitating a desire that is itself imitating one's own desire, as we have seen above. There is no original desire, and the object on which rival desires converge is the *emerging* production of the mechanism itself. (Think of two absent-minded professors going together to attend the same event. Neither of them knows the venue; each one believes the other knows. A trajectory emerges, endowed with some stability, from the fact that each partner follows in the other's footsteps.)[22]

Let's return to the Polya's urn scheme. There is a fundamental difference with the Buffon's needle case. Every time the experiment is carried out again, the same phenomenology obtains, but the value towards which the dynamics converges is different. It is entirely contingent on the experiment in question. The dynamics seems to be converging towards a preexisting value, but the value is generated by the very dynamics. From within, it is impossible to realize that a preexisting end does not guide the evolution. From without—that is, if we are able to pull ourselves up by our own bootstraps and contemplate from there the set of all possible trajectories—our specific world appears in all its contingent singularity to be one among a manifold of possibilities. It is a case of "complexity from noise." Chance here brings about a form of necessity that appears as such only retrospectively.

The relationship between the dynamics and its asymptotic behavior (called, in the jargon of the mathematical theory of dynamical systems, an *attractor*) takes on the form of a loop that is the signature feature of a self-organizing system—that is, a loop between an emerging level (the attractor) and its conditions of production (the dynamics):

EMERGENCE

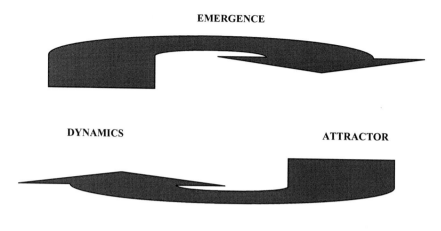

DYNAMICS   **ATTRACTOR**

CONVERGENCE

Figure 1. *Complexity from noise*: the dynamics converges towards an attractor that is generated by itself. The evolution is said to be ***path-dependent***.

Over the years, I have shown that all the mimetic and sacrificial figures that MT has put forward or brought out (double mediation, pseudo-narcissism, pseudo-masochism, the pattern of the "Stranger," the scapegoating mechanism) are particular instances of this bootstrapping scheme.

This scheme is consonant with an impressive series of scientific and mathematical discoveries made during the second half of the twentieth century that have completely changed the way in which we conceive the trajectory of a material system subject to purely causal physical laws. It is well known today that *complex systems*, made up of many elements interacting in nonlinear ways, possess remarkable properties—so-called emergent properties—that justify their description in terms that one should have thought had been forever banished from science in the wake of the Galilean-Newtonian revolution. Thus it is said of these systems that they are endowed with "autonomy," that they are "self-organizing," that their paths "tend" toward "attractors," that they are "path-dependent," that they have "intentionality" and "directionality"—as if their paths were guided by an end that gives meaning and direction to them even though it has not yet been reached; as if, to borrow Aristotelian categories, purely efficient causes were capable of producing effects that mimic the effects of a final cause.

In a sense, we are not far here from Kant's conception of nature in the second part of his third critique, the *Kritik der Urteilskraft*, entitled "Critique of Teleological Judgment."[23] Only explanations that ultimately appeal to causal mechanisms are considered adequate. Nonetheless, faced with the most surprising manifestations of complexity in nature (life, for Kant), recourse to another "maxim of judgment"—teleological judgment—becomes inevitable. Concepts such as "internal finality" are indispensable, and perfectly legitimate, so long as one keeps in mind that they have only heuristic and descriptive relevance. Teleological judgment consists in treating them as though—the Kantian *als ob*—they have objective value.

An essential feature of MT, which unfortunately is overlooked or misunderstood by many who are exposed to it, is that all the accounts it provides are entirely causalist, as befits a scientific theory. And nevertheless, they are capable of explaining why human history, in spite of its fury and apparent madness, is not a tale told by an idiot, but displays features that evoke intentionality and directionality. The genesis of the scapegoating mechanism—to take up the figure that is at the core of Girard's anthropology of violence and the sacred, but also one about which radical misunderstandings abound—was not invented by humankind *in order to* keep its violence in check. It constitutes one possible *attractor* of the dynamics of violence.

Mimesis and Cultural Evolution

The Polya's urn model is the simplest model capable of generating "complexity from noise," in contradistinction to "order from noise." Much work has been carried out in recent years to enrich it in various ways. It is remarkable that a good many of the models that have thus been developed can be subsumed under the category of "cultural evolution." It is no less outstanding that all of them place the logic of imitation (under various guises) at center stage.

A very active branch of formal economics is today exploring the role of what it calls "interpersonal influences" in economic activity, a euphemism for mimesis. Economists are progressively realizing with surprise and reluctance how far removed this mimetic universe is from the ideal market. Contrary to what they thought, generalized imitation produces something rather than nothing—a statement that constitutes the "b a ba" of MT! It creates self-reinforcing dynamics that converge so resolutely on their target that it is difficult to believe that this convergence is not the manifestation of an underlying necessity, in the manner of a mechanical or thermodynamic system returning invariably to its equilibrium state after straying from it under the effect of some perturbation. Yet one sees that the concept of equilibrium, which the theory of the market imported from rational mechanics, is absolutely unsuited to characterize the "attractors" of mimetic dynamics. Far from expressing an implicit order, they spring from the amplification of an initial fluctuation, and their appearance of preestablished harmony is a mere effect of unanimous polarization. They are condensations of order and disorder. The mimetic dynamic seems to be guided by an end that preexists it—and that is how it is experienced from the inside—but it is in reality the dynamic itself that brings forth its own end. Perfectly arbitrary and indeterminate a priori, it acquires a quality of self-evidence as the vise of collective opinion tightens. If there is a social process that illustrates to the highest degree the notion, so important today in social and political philosophy, of "pure procedure," it is the mimetic dynamic. There is no other way to determine its result than to let it proceed to its conclusion. It is a random procedure that takes on an aura of necessity.

In coming to an equilibrium, the economists' ideal market is supposed to reflect an external reality. The prices express objective, "fundamental" values that synthesize information as diverse as the availability of techniques, the scarcity of resources, or the preferences of consumers. The mimetic dynamic for its part is completely closed upon itself. The attractors that it generates are not in any relationship of correspondence with an external reality; they simply reflect a condition of internal consistency: the correspondence between a priori beliefs and a posteriori results. The mimetic attractors are self-realizing representations.

Generalized imitation has the power to create worlds that are perfectly disconnected from reality: at once orderly, stable, and totally illusory. It is this "mythopoetic" capacity that makes it so fascinating. If there are hidden truths somewhere to be discovered, one must not count on mimetic dynamics to disclose them. If it is real-world efficiency one is looking for, it is again better not to have to depend on them. Efficiency and the capacity to reveal hidden information: those are two properties that economists readily attribute to the ideal market. The distance between the latter and the mimetic process seems insuperable.

The clinical picture of the imitative logic is in its essentials already present at the stage of a very simple model in which the mimetic connections between agents are given and remain fixed throughout the whole process: the probability that a given agent imitates another given agent is a constant, possibly null.[24] Phenomenologically, we know that this hypothesis is too restrictive and that the mimetic dynamic has the ability to modify the structure of its own connections: one subject has all the more chances of being imitated by another given subject if he is already imitated by many other subjects. An opinion's power of attraction increases with the number of individuals who share it. One can see that if this is the case, the effects of mimetic polarization are accentuated accordingly. It may seem, however, that such hypotheses depend too much on the irrationality of crowd phenomena. In fact, research in recent years has shown that they correspond to forms of behavior that are individually rational. For instance, there are cases where the personal advantage that an individual derives from joining the mass grows *objectively* with the size of the latter.

This hypothesis is today precisely a commonplace in the literature that deals with a very important dimension of cultural evolution in modern societies, the *choice of techniques*.[25] As a technique spreads, more is learned about it, and it develops and improves. The more users there are, the richer and more diversified the selection of products becomes; production costs diminish, and so does the risk of failure. In these conditions, the competition between rival techniques displays features that distinguish it markedly from the "perfect competition" of economists. The first is the multiplicity of attractors. The selection of one among them cannot be determined by deduction from the formal structure of the problem; it is the actual history of events, with its contingencies, fluctuations, and random turns—especially those affecting the system's first steps—that are responsible. The concept of "path-dependency," which the Polya's urn scheme illustrates so simply and cogently, plays a crucial role here. Fundamentally, it expresses the same idea as that of "pure procedure." We are poles away from Le Chatelier's principle, a thermodynamic reference still popular with theorists of the market who want to laud the latter's capacity to neutralize perturbations that affect it. The evolution of such a dynamic is highly unpredictable. There is obviously no reason for the selection that it accomplishes to be the most efficient one. If a certain technique is favored by chance at the outset, it will benefit from a "selective advantage" that it will maintain and amplify as the number of users grows. It may end up dominating the market, even though another technique would have shown itself to be more advantageous for everyone if only chance had selected it from the start. Technological evolution thus has a strong propensity to get locked into undesirable paths from which it is harder and harder to be removed. Chance, selection, "order through fluctuation," self-organizing process: all of these terms used today by historians of technology define a theory of evolution that has only the remotest kinship with neo-Darwinism, but owes all of its most stunning characteristics to the logic of imitation.

Any theory of evolution, in biology or elsewhere, that is structured by the "order from noise" model suffers from a fatal flaw: it is unable to account for the diversity of the world. The dynamics of the latter are bound to converge towards preexisting states. ET, in biology (or worse, in the social sciences), when it refers to a principle of selection such as the "survival of the fittest," falls under that fundamental critique. Darwin himself saw the danger. He did not think that selection, the sacred cow of ET, was the only factor in evolution. Today, one could even say that it is not even the most important one. In the first edition of *On the Origin of Species*, Darwin wrote: "I am convinced that selection has been the main, *but not the exclusive*, means of modification." In the sixth edition, he added: "Ever since the first edition, I have repeated this claim again and again. This has been of no avail. Great is the power of misrepresentation."[26]

In order to escape that fate, mechanisms capable of generating "complexity from noise" are required. Mimesis is the ground from which they stem.[27] That is why MT gives the impression that it can account for everything, from quantum to neuron to the eschaton. This impression is not unwarranted.

NOTES

1. This term was coined by Robert Hamerton-Kelly during one of our meetings at Stanford.

2. Walter Benjamin, *Reflections: Essays, Aphorisms, Autobiographical Writings*, ed. Peter Demetz, trans. Edmund Jephcott (New York: Harcourt Brace Jovanovich, 1979), 333.

3. For example, see Ann Cale Kruger, "Imitation, Communion, and Culture," Vittorio Gallese, "The Two Sides of Mimesis: Mimetic Theory, Embodied Simulation, and Social Identification," and Andrew N. Meltzoff, "Out of the Mouths of Babes: Imitation, Gaze, and Intentions in Infant Research—the 'Like Me' Framework," all in this volume.

4. Jean-Pierre Dupuy, *The Mechanization of the Mind* (Princeton, NJ: Princeton University Press, 2000); Jean-Pierre Dupuy, *On the Origins of Cognitive Science* (Cambridge, MA: MIT Press, 2009).

5. Analytic philosophy of mind is divided today between two camps, one named "Theory Theory" (TT), the other "Simulation Theory" (ST). TT has it that people read other people's minds by attributing to them mental states from which they derive explanations and predictions regarding their behavior. By contrast, ST posits that the observer uses his *own* mental mechanisms to predict the mental processes of others. The discovery of mirror neurons has given a boost to ST. It is not surprising that the major founder of ST, philosopher Alvin Goldman, collaborated with one of the discoverers of mirror neurons, Vittorio Gallese, to draw out the philosophical implications of this discovery. Their joint article "Mirror Neurons and the Simulation Theory of Mind-Reading" (*Trends in Cognitive Sciences* 2, no. 12 [December 1998]: 493–501) has become a classic. In it we can read: "Mind-reading is the activity of representing specific mental states of others, for example, their perceptions, goals, beliefs, expectations, and the like. It is now agreed that all normal humans develop the capacity to represent mental states in others, a system of representation often called folk psychology. ... The hypothesis explored here is that mirror neurons

are part of—albeit perhaps a rudimentary part of—the folk psychologizing mechanism" (496). What is remarkable in this quote is that ST, no less than TT, claims to provide a scientific, physicalist account of folk psychology. Gallese has since revised his position (see his chapter, "The Two Sides of Mimesis: Mimetic Theory, Embodied Simulation and Social Identification," in this volume), and his more current understanding of "embodied simulation" more closely approximates the implicit theory of mind found in the mimetic theory. The fact that MT has started a collaboration with Vittorio Gallese, as testified by this volume, is significant.

6. Bertrand Russell, "The Philosophy of Logical Atomism," in *Logic and Knowledge: Essays 1901–1950*, ed. Robert Charles Marsh (London: Unwin Hyman, 1956), 227.

7. Donald Davidson, *Essays on Actions and Events* (New York: Oxford University Press, 1980).

8. Pascal Engel, *Introduction à la théorie de l'esprit* (Paris: La Découverte, 1994), 10.

9. René Girard, *Deceit, Desire, and the Novel*, trans. Yvonne Freccero (Baltimore: Johns Hopkins University Press, 1966), 99.

10. René Girard, *Things Hidden since the Foundation of the World*, trans. Stephen Bann and Michael Metteer (Stanford, CA: Stanford University Press, 1987), 370–71.

11. That is what I have set out to show in my *The Mechanization of the Mind*, trans. M. B. DeBevoise (Cambridge, MA: MIT Press, 2009).

12. Daniel Dennett, *The Intentional Stance* (Cambridge, MA: MIT Press, 1987).

13. Francisco Varela, "Organism: A Meshwork of Selfless Selves," *Rev. Europ. Sciences Soc.* 29 (1991): 173–98.

14. Daniel Dennett, "The Origin of Selves," *Cogito* 3 (Autumn 1989): 166.

15. Marvin Minsky, *The Society of Mind* (New York: Simon and Schuster, 1988).

16. Antonio Damasio, *Descartes' Error* (New York: G. P. Putnam's Sons, 1994).

17. Friedrich Hayek, *Law, Legislation and Liberty*, vol. 1, *Rules and Order* (London: Routledge & Kegan Paul, 1973), 22–23.

18. Ibid., 23.

19. Those phrases have been coined by the neo-cybernetician tradition, from Heinz von Foerster to Francisco Varela to Henri Atlan. A history of those concepts can be found in Dupuy, *On the Origins of Cognitive Science*.

20. One hundred years before Darwin, Buffon, in his *Histoire Naturelle*, postulated the existence of a common ancestry of man and apes.

21. George Polya, who died in 1985 in Palo Alto, California, at the age of 98, was the professor of the mathematician John von Neumann. Neumann, to whom we owe the concept of complexity, was also of Hungarian origin.

22. An Argentinian ad for a business magazine illustrates this mechanism in a brilliantly humorous fashion. Two blind men are about to cross a street. The dialogue between them goes like this: "Cruza?" "Si." ("Are you crossing the street?" "Yes.") And the two set out to cross the street, each one giving his arm to the other, in the middle of a heavy traffic that almost runs them over. Neither of them understands that the other is blind. The first one is really asking: "You see that I am blind. Would you help me cross the street?" The second is answering yes to a different question: "I see that you are blind: do you want me to help

you cross the street?" The lesson of the ad: it is better not to be informed than to wrongly believe that one is. The recent financial crisis was driven in part by mechanisms of this kind.

23. Kant, "Critique of Teleological Judgment," *The Critique of Judgement*, trans. J. H. Bernard (London: MacMillan and Co., 1914), 259–429.

24. See the model presented by André Orléan, "Money and Mimetic Speculation," in *Violence and Truth*, ed. P. Dumouchel (Stanford, CA: Stanford University Press, 1988), 101–12. Under certain conditions, it is demonstrated that the imitative dynamic converges toward unanimity of the group. These conditions reflect the fact that there is an effective interdependence among all the agents; in other words, very few probabilities p_{ij} are null, p_{ij} being the probability that the agent i imitates the agent j.

25. Brian Arthur, "Competing Technologies: An Overview," in *Technical Change and Economic Theory*, ed. G. Dosi et al. (London: Pinter Publishers, 1988), 590–607.

26. Darwin, *The Origin of Species and the Descent of Man* (New York: The Modern Library, 1936), 367.

27. I venture the following conjecture, although I am quite unable at the moment to justify it: Mimesis, in one form or another, is not only sufficient, but also necessary for the generation of complexity from noise.

Mimesis and Science: An Interview with René Girard

The following interview is composed of material collected over the course of this book project, including a two-day interview conducted by Scott Garrels (S.G.) with René Girard (R.G.) at his home in Stanford, California, on July 18–19, 2008. Additional material was taken from presentations given by Girard at the project meetings at Stanford during 2007–2008, including questions that were asked of him at the time by several of the authors in this volume. For the sake of consistency, the single interviewer S.G. is used throughout.

Part 1

The Mimetic Theory: An Overview

THE HISTORY OF IMITATION

S.G.: I want to begin by discussing your Mimetic Theory as a whole before asking you some questions about the recent empirical research. This volume is an attempt to build a bridge between the new sciences of imitation and your work on mimesis and violence, which is more relevant now than ever.

R.G.: If my theory has convinced you of that, then I'll take it as a sign of success, since many people have dismissed the notion of imitation from the very beginning. I remember the reaction to my first book, *Deceit, Desire, and the Novel*.[1] I was told that I had written a very creative book, but that my use of imitation was like the proverbial hair in the bowl of soup. People asked: "What does it have to do with anything? Imitation is not interesting." I always answered, "I'm sorry, but it's absolutely fundamental to what I am saying."

S.G.: In the 1950s and '60s imitation was not a very popular topic.

R.G.: That's true. Even so, I was surprised that my ideas did not catch on sooner. Perhaps researchers were prevented from going to them because they were too commonsensical. It is clear that all human relations are based on imitation. The worst, like the best of them. If someone does something kind to you, you do the same; you are compelled to do the same. If you don't, something is wrong. Therefore, you imitate them. And if they start being mean to you or turning their back on you, you turn your back too. You manage to make it known to them that you understand how they feel about you and that you feel the same. This often means that you add a tiny bit of disagreeableness to the existing disagreeableness as you see it. However, this little something added is going to look to the other like an enormous provocation, like a declaration of war. From there, your relations are going to go from bad to worse. But whether you exchange compliments, greetings, insinuations, signs of indifference, meanness, bullets, atom bombs . . . it's always imitation.

Imitation is everywhere. It was addressed by the greatest of all philosophers, Plato. Aristotle, too, defined man as the most mimetic of all animals, which is a very profound definition and still of value today. And by the end of the nineteenth century, there was a period where imitation was supposed to explain everything. The nineteenth-century sociologist Gabriel Tarde wrote a book called *The Laws of Imitation*, which is still revived from time to time.[2] So before our period, imitation was extremely fashionable. At the same time, the theoreticians took all the drama out of imitation. Tarde, for example, didn't see the negative aspects of imitation. He didn't see the rivalry. He didn't see that imitation is the main source of violence in humans.

By the twentieth century, imitation was rejected precisely because it did not incorporate these other aspects. It seemed too facile and could not account for the wide range of phenomena that many were attempting to explain. When you say "imitation," everybody thinks of being sheep-like, gregarious, following people, and so forth. This is true in many instances, but what is also true is that imitation not only affects your gestures, your words, or your ideas; you also imitate desires.

It's surprising in a way that the observation is not more common. We desire something because others find it desirable. So things like friendship or discipleship are highly susceptible to conflict because they are based on openness to the other, to their desires in particular, which then bring people into conflict over something that cannot be shared; this in turn can create competition and all sorts of drama. Usually students are well aware of this. I would say to my students at Stanford: The fact that you are friends gives you plenty of opportunities to be enemies; you are friends because you follow the same path in life, but this means that sooner or later you may compete for the same fellowship or the same job or the same boyfriend or girlfriend. Is your friendship going to hold up? You should be ready for that test. This rivalry is obviously the greatest source of conflict between nations. They are rivals for the land that apparently belongs to them both. Nationalism is about claiming territory that the neighboring country desires. And you will invent that territory, that object of conflict if it doesn't exist already. Just look at France and Germany fighting over Alsace in the nineteenth century. And of course there are many such examples today.

MIMETIC DESIRE IN LITERATURE

S.G.: You began exploring imitation while teaching and studying literature, where you found its relationship to human conflict represented in works by such authors as Cervantes, Dostoevsky, Proust, and Shakespeare.

R.G.: Yes, the first thing I discovered is that human conflict is complicated because it is linked to imitation, and the only people who really know that are the great writers. Shakespeare talks about nothing else.

Most of the first comedies of Shakespeare begin with two young friends, sometimes four, as in *A Midsummer Night's Dream*. They have been friends since early childhood. They are friends (they say so themselves) because they do and like all the same things. They like the same books, the same wine, and so forth, and then suddenly they fall in love with the same girl. And then all hell breaks loose.

Shakespeare comes back to this theme at the end of his life in the greatest of his plays, *The Winter's Tale*, which like *Othello* is a play about jealousy—in this case, the insane jealousy of a man for his best friend. If you love the same thing as your best friend, he becomes your best enemy. The negative and the positive are in strong conjunction. In Shakespeare there are many, many lines which express the mystery of conflation of the greatest affection for your friend and the greatest jealousy for the man who is in love with the same thing. The human sciences should assimilate this insight. They have not.

S.G.: What books initially led you to these discoveries about mimetic desire and conflict?

R.G.: Two books from my childhood, first of all. I talked about one of them, *Don Quixote*, quite a bit in my first book. The other one is *The Jungle Book* by Rudyard Kipling. I've never talked about the influence of *The Jungle Book* until recently. It is an incredible book of animal stories, which are also excellent mimetic stories. My main insight came when I encountered *The Jungle Book* and *Don Quixote* along with the writings of Proust.

Proust's *Remembrance of Things Past* is more revealing of mimetic desire than any other modern book. I'll take just one example. As a child, the main character is very precocious in his love of art, and in particular of theater. At some point, his parents decide to send him to the theater because he wants to see a famous actress, La Berma (who is based on Sarah Bernhardt), playing the lead role in Racine's *Phèdre*. He fell in love with La Berma mostly through what one could call the mimesis of writing; he saw the advertisements for the theater in the streets of Paris. Reading about La Berma every day when he was going to the Champs-Élysées on his walks awoke in him a passion for La Berma, although he had never seen her. But he had heard people celebrate her. So he goes to the theater and watches La Berma. He's terribly disappointed. It's reality after the dream. He doesn't find anything interesting about her or her performance, and he leaves completely disenchanted.

But the following day, an old diplomat, who is a friend of his father, comes to the house. Norpois is an old fool and a complete nonentity. The young boy is 50 times more intelligent and sensitive about art than he is. The father says, indicating his son: "Yesterday, he went to see La Berma." Norpois speaks a few words of praise for La Berma. It's pure politeness and courtesy and means absolutely nothing. But it is enough. The voice of the old diplomat has so much unspoken authority that it restores the child's belief in La Berma. Retrospectively he starts to enjoy the performance, even though in reality it bored him to death. His faith is restored through purely mimetic means.

The amazing thing is that Proust's writing gives an impression of personal intimacy, of absolute faithfulness to the reality of desire, which the critics describe as such. But they never tell you that the source of it is Norpois, who has nothing to do with the child but is nevertheless impressive to him because he's a famous man. In other words, what is most exterior becomes inner experience. If you read Proust naively, you will talk about inner experience without seeing the imitation at all. Imitation makes no sense in normal psychological terms. But normal psychology is the one Proust describes, not the one we imagine. Imitation is what children do. The spontaneity of imitation is their virtue, and this scene is one of the most exemplary in all of Proust's works.

In order to understand the power of the scene, you must compare it with Proust's first novel. The first volume of *Remembrance of Things Past* was published just before World War I. But years earlier, Proust had written a book which instead of being in the first person, is in the third person. The hero is named Jean Santeuil. He is a perfect young man who believes in all the right things. He's very happy with his friends. He never has the slightest trouble with anything. There is a theater scene in this novel too. The beautiful people are there in the boxes, and the hero is with them. A former king of Portugal even helps him to arrange his necktie. The young man has all the success in the world. He never fails. And this first book is no good at all. It's incredibly boring, uninteresting, and insignificant.

In the second book of *Remembrance of Things Past*, the same scene is reinterpreted. This time the hero is sitting down below in the orchestra seats. And up there in the loge is the Duchesse de Guermantes, this noble lady with whom he's in love from a social viewpoint. What he wants, really, is to be recognized by her. And he's looking at her from down below, in the orchestra seats, and seeing her in that loge, which is like some kind of paradise compared to the hell of the pit he's in. The Guermantes pay no attention to him. They don't even see him. The difference is that in this second book he can put himself in a bad position. We are normally so egotistical that we cannot place ourselves in the unenviable position. We have to brag even when we invent fiction. The great novelist is someone who stops doing that. And this fall from grace is usually situated at the end of the book. In Proust, the book is divided into two parts; first, the illusions of the narrator, and then reality. The two are separated by disillusionment. The Spaniards say *desengaño*.

In Proust, the fall of the narrator is in the last volume of *Remembrance of Things Past: Time Recaptured*. The last volume is the origin of the novel. It describes the realization that one has forever surrendered to mimetic incitement in one's own life. And the result is the writer's ability to represent that mimetic incitement, to which he always surrendered—to represent it truthfully, something he couldn't do in his first book, which was pure bragging, denial of reality. If you look at *Time Recaptured*, which is in my view the place where the narrator talks about the genesis of the final novel, he doesn't talk about the first book. Or if he does, it's only to mention that he tried to write and it didn't work. But the final volume is the beginning of the real novel. There's a distinction between *Time Lost* in Proust, the time re-created in the novel, and *Time Recaptured*, which enables him to write about it. The experience is present in the novel.

Proust was a very successful snob. He frequented the best circles of the Faubourg St.-Germain, the old aristocratic part of Paris. But eventually he became aware that his social climbing was a totally metaphysical thing, because these people had no power. They had a little money left perhaps, but not as much as the industrialists in the early 1900s. Their time was past. By comparison, the snobbery described in the French novel of the nineteenth century is more concrete. In the middle of the century, when the earlier French novelist Balzac was writing, the aristocracy was still rich and powerful enough for social climbing to make sense in a concrete sociological way. When Proust is writing, it doesn't make sense any more. And the genius of Proust is to understand that. The people who reproach Proust for writing about this aristocracy don't understand that thanks to historical circumstances, he wrote a work about the quintessence of snobbery, which has nothing to do with reality and is purely metaphysical, "religious." The nonsense that is the object of desire should not be grounds for accusation. Proust is the one who created that nonsense in order to show it to us. Mediocre writers of the same period talk about snobs as if they still amounted to something. The people who condemn Proust for writing about snobs don't understand his genius. He shows their nonentity in the most concrete fashion. His entire book is devoted to doing that. This is the first book I understood from the mimetic perspective. Comparing it with *Don Quixote*, I realized that Cervantes talks about the old chivalry just as Proust talked about the snobs of the Faubourg St.-Germain in the early twentieth century. When I realized that, I had my first book. It was just a question of showing that there were differences in the centuries, in line with the historical and sociological changes that were happening.

In a way, all novels are about "snobbery," which is just religiousness displaced onto a social object of little significance. But precisely because many readers attach much more significance to this nonsensical object than do the writers, their genius goes unrecognized. It is misunderstood by people who think that they are better than the writer because they condemn something that is too insignificant to be worth condemning. To treat this insignificance in a truthful way is an enormous achievement. After all, every one of us has some insignificance with which he is tremendously in love. In the right circumstances, if you can conquer one, you can conquer all other such obsessions.

S.G.: In *Deceit, Desire, and the Novel*, you speak of the novelist undergoing a "conversion" from blindness to his or her own mimetic snobbery or hypocrisy, to an understanding of the role that imitation plays in his or her life. This conversion is the Proustian experience you have just explained, the one that makes the writer aware enough to be able to write a masterpiece.

R.G.: Yes, in *Deceit, Desire, and the Novel* I used the word "conversion" as a technical word. It denotes a change of attitude, one that normally doesn't take place because your mind is used to reacting in a certain way. But some experiences may trigger a real change. To convert means to turn around and do something differently. No more than that. Something is triggered because a certain change has taken place or was ready to take place in you. Novelists are really aware of this. They say that the novel is the product of such a change. Proust quotes the conclusion of novels that imply the thing that he states explicitly. For instance he talks about the conclusion of *The Red and the Black* by Stendhal, in which this change happens in the hero, Julien Sorel. Proust says it's really the novelist who undergoes that change. Proust makes it possible to see the connection between all novelistic conclusions, in which the drama becomes a symbol of the writer's internal change. So I didn't make it up. The idea came from reading these conclusions.

MIMETIC VIOLENCE

S.G.: In the 1960s, after completing your first book on mimetic desire, which also included important elements of mimetic rivalry and conflict, you became interested in questions that took you well beyond literature, into the realm of anthropology and ethology.

R.G.: Yes, I began to wonder about what differentiates humans in the animal kingdom. I asked: when does humanity begin? We know that animals already engage in mimetic rivalry. And that mimetic rivalry in animals is the source of what biologists today call their culture. When two animals desire the same female or the same territory, they fight over them—but they do not fight to the finish. They fight moderately, and the weaker animal acknowledges his weakness and the stronger animal spares his life. The stronger becomes the dominant animal and the other one the dominated animal. It ends without death—at least that was the consensus when I was doing my research for

Violence and the Sacred.[3] The compromise in animal society was called dominance patterns. Animal culture was based on hierarchies of dominance. There was absolutely no possibility of death, it was thought. There were moments when chimpanzees would have complete chaos in their community, but, according to the research at the time, they didn't kill each other. It was known that chimpanzees killed other monkeys, but not that they also killed their own kind.

S.G.: You are referring to more recent field observations of male chimpanzees forming a pack and going out and killing other isolated chimpanzees.[4]

R.G.: Yes, so perhaps it is now known that chimpanzees do kill each other. But regardless, we know that a human being will fight to the finish, will kill his opponent, his rival. Not only that, human beings will fight beyond the finish. If you fight to the death with someone in an archaic culture, that's not the end of it. The victim's brother will pick up the fight, and kill the murderer. And then the brother of the second victim will pick up the fight and so forth. But if this happens, of course, there cannot be any dominance factor; there can be only continuous fighting, people being killed. So human beings have vengeance, which is worse than killing.

S.G.: Vengeance is a form of mimetic violence?

R.G.: Vengeance is absolutely mimetic violence. It's doing exactly the same thing to the other that he's done, not to me, but to another person, maybe a family member. Revenge is not a cultural institution. If you start looking at archaic culture you will see that revenge is pretty much universal. It's what we call the "blood feud." And the blood feud is already a religion of violence; it transcends individuals, it transcends time, it may transcend space. If your relatives go on endlessly taking up the fight with the last murderer, society is sure to end right there. The whole community becomes involved, and then it's complete chaos. Everybody is fighting everybody else. So the question I asked is: how can humans form societies, long-term associations, given the existence of that form of rivalry, which is endless and goes on forever?

The Scapegoat Mechanism

S.G.: Your answer to that question was the discovery that collective violence deflected against a common victim puts an end to the back and forth of violent reprisals. What is the tipping point in the evolutionary process from nonhuman primates to humans that brings about that transition from vengeance to the scapegoat mechanism your theory describes?

R.G.: The key is that mimesis increased over time in proto-humans. In the course of fighting, what must have happened at the time of hominization, over hundreds of thousands of years maybe, is that mimetic rivalry became so intense that the dominance patterns somehow disappeared. When dominance patterns no longer take hold and you are fighting mimetically for the same object of desire, you cannot be reconciled. Sooner or later the object will be destroyed or forgotten or become less significant in the course of fighting, and only antagonists will be left. At that point, you have a community which is in a very serious crisis and which is mimetically mad. Each member is imitating each other's violence. But the strange thing is, when you only have antagonists left, you can share an antagonist. You can become aligned with another person against a third party. Mimesis, which opposes people when they desire the same object, suddenly joins them together. Two antagonists against one will have more mimetic power, and their side is going to attract more and more people. Ultimately there will be total imbalance. All antagonists will be on one side against a single person. When this happens, if this happens, and even if others are killed in the process, something different emerges in humans when all join collectively against the same antagonist. This antagonist becomes the original scapegoat, the single victim. When that single victim is killed, all the antagonists find themselves without any antagonists; therefore, they are de facto reconciled.

S.G.: Would you say this is a random event?

R.G.: You can never say it is going to happen for sure. It may happen once in every ten times, and nine communities may be destroyed. But the shift from the mimesis of fighting over objects to the mimesis of fighting against someone always becomes cumulative in the same way. The key to the scapegoat is very simple. How could a crisis be reduced to a single victim? Ultimately, with that kind of imitation, the contagion of antagonism, there will be only a few victims left, and then finally one. When that one enemy is killed, there are no enemies left in the community, and peace returns. That is why the victim becomes the god. The victim is regarded not only as very bad, very dangerous, because it made us fight, but also as very good, because it reconciled us. That single victim is the origin of the archaic gods, who are very violent but also very peaceful when they want to be. They are both at the same time.

S.G.: Your theory emphasizes that religion was there at the beginning of humanity. That it's not something that was made up after human culture got off the ground. Religion emerged as an adaptive solution to very real and unprecedented social problems that were a result of the natural course of evolution in the earliest stages of hominization.

R.G.: When people talk about religion today, or against religion, they tell you that religion is a view of the world. In fact, most scientists are stuck on the idea that religion is an outdated view of the world, a fanciful view of the world that arises because archaic people didn't know how to explain the mysteries of the universe. But archaic societies couldn't care less about the mysteries of the universe. They never talk about them. They don't even suspect there are mysteries of the universe. You have to be incredibly sophisticated to look at the starry night and invent stories. That's the way the nineteenth century saw the beginning of religion: people look at the stars and they invent the gods. Not true at all. They have more serious problems to contend with, which are not only problems with the outside world, the forces of nature, disease and so forth, but violence inside the community. The imitation of violence gets worse and worse as we get closer and closer to humanity. Therefore at the very moment when humanity needs the most protection against its own violence, suddenly there are no dominance patterns to do the job. So what protected us instead? A lynching.

In my view, this is the origin of archaic religion. The scapegoat reconciles the whole community. It's a Darwinian genesis of humanity. The shift from animal to human cannot be explained simply by organic changes. The invention of humanity is the invention of culture, and the invention of culture is also the invention of religion. There is no archaic culture that is not religious. There is no archaic religion that is not the reinvention of culture. You find that sacrifice is at the heart of all archaic communities. It's very weird, I admit, but read the archeological literature. Around 1860 or before, we have anthropologists, most of them English, who recorded religious institutions of archaic religions in a manner that is incomparable. And my theory is a result of the reading of these English anthropologists. I think you can unify the whole thing and find behind it the same phenomenon, which is always interpreted a little bit differently by each community. But it always turns into a religion. The early anthropologists were very close to discovering that. I think they just didn't dare say that violence, conquered by sacrifice, might be the origin of religion and of culture as a whole.

S.G.: What then is sacrifice, ritual sacrifice?

R.G.: If human communities have an experience like the one I just described, at first they are very happy about how they are all reconciled. They all kiss each other, they love each other again and are very grateful, but they are still human, and they will inevitably become rivals again. And when this happens, what will they do? They will remember that in the past there was a single victim that reconciled them. And they will mimetically recreate the resolution with substitute victims, and ritual sacrifice in my view is nothing else.

When the anthropologists of the late nineteenth century would talk to primitive tribes, they all talked about their sacrifices in the same way—a god gave them to us, and this god gave them in order to keep peace among us. Peace and the gods are always mentioned, and I believe it's true. This victim's not really dead, since he saved us. After destroying us halfway, he saved us. He brought us peace. And peace becomes so important in that crisis that you worship whoever brings it to you. The supposedly most violent creature, as it turns out. The proof that they want to repeat this collective lynching, which is really a grand and unprecedented phenomenon, is that they sacrifice someone, and they do so in a way that resembles a sort of mock crisis at the beginning of the ritual.

RITUALS AND PROHIBITIONS

S.G.: How do you account for the complexity and diversity of culture and religion as emanating from this same phenomenon, a collective murder?

R.G.: You can see that archaic religions are divided into three parts, which seem contradictory but are nonetheless related to the same phenomenon. The first part is prohibitions: what you shouldn't do, in order to prevent mimetic rivalry and violence. The second part is sacrifice, which is very strange because in some ways sacrifice is the opposite of prohibitions. Prohibitions forbid violence, whereas sacrifice requires of you one violent action, which is the killing of a victim. But you discover that this peace does not last very long, and as a result, sacrifice must be repeated more frequently and predictably. This is ritual, which brings you back to the situation that you had with the initial de facto reconciliation. Prohibition is first: let's not do it. If the prohibition doesn't work, ritual comes into play: let's do it again, but let's decide ahead of time who the victim will be so that we do not fight about it. That's what sacrifice is—to be sure that we all decide on the same victim ahead of time. So you have that mystery of religious sacrifice and prohibitions, which are at the same time against violence.

S.G.: Prohibition gets a bad rap in the contemporary world, particularly in popular culture, and especially when it comes to religion. We view prohibitions as essentially an obstacle to fulfilling our desires, or an abuse of power for social control. But you're saying that religious prohibitions, especially those in archaic cultures that to us might appear rather absurd, had an essentially life-saving function.

R.G.: Prohibition has only one object: the prevention of violence inside the community. Today it's very fashionable to denigrate prohibitions, which are seen as completely irrational. But archaic prohibitions are completely rational. Sometimes, however, archaic people have a conception of violence that is just empirically flawed. They are aware that the more people are alike, the more they fight—and this is why so many archaic cultures are against twins. There are many archaic cultures that understand that twins have nothing to do with violence, and they pay no more attention to twins than we do. But there are others that will not tolerate twins. They feel that if they allow twins to exist, violence will spread like wildfire and destroy the whole community. They don't kill twins out of meanness of spirit. They simply think that the birth of twins has something to do with violence. They often think that the mother has been misbehaving or transgressing prohibitions against violence, and that if you transgress prohibitions against violence you'll have twins. In other words, that you'll produce mimetic rivalry, "monstrous doubles." Many mythical heroes are twins. Why? Because twins fight all the time. Or are supposed to. And they kill each other. Romulus and Remus are two twin brothers. Cain and Abel are not far from that. Many communities, I repeat, understand it's not true in the case of biological twins, but many don't. And as a result, sometimes they will get rid of only one twin, which shows that it's the similarity of twins that bothers them.

S.G.: What is the origin of the particular rituals and their diversity that are widespread in all human cultures and apparently even among Neanderthals, like the ritual of burying the dead?

R.G.: In my view, rituals stem from the original scapegoat, which founds the community. There can be countless variations. Sacrifice intervenes on occasions where there is often the possibility of trouble being stirred among members of the community: birth, the death of a relative, and so forth. The rituals exist because the culture tries to purge itself ahead of time of any tension or rivalry. The more sacrificial victims we kill ahead of time, the more we create a bond between us, and the less chance there is that we're going to fight each other.

Another example would be rites of passage for male youth. Rites of passage are virtual sacrifices. They involve an ordeal that reproduces the crisis we have been talking about. Sacrificial rituals are nothing but the most accurate repetition possible of the spontaneous scapegoat phenomenon. They are a precaution against possible trouble, in anticipation of it. The tendency is to multiply rituals ad infinitum in order to make peace, but at some point the more you perform them, the less efficient rituals are.

S.G.: Can you say something about the cathartic aspect of these rituals?

R.G.: Rituals are supposed to repeat the effect of the first murder. We have one description of that effect and it comes from Aristotle in the *Poetics*. He calls it "catharsis," which means purification—purification of violence. There are many archaic myths in which two groups are fighting together until they discover that there is a malevolent individual who pretended to belong to one group or who was casting stones at the other group. When they can agree on that one individual, they are saved. And that's what catharsis fundamentally is—purification of human relations.

S.G.: How do you explain the convenience that a guilty victim is always on hand when the community is most in need of one?

R.G.: When we use the word "scapegoat," we think about an entire group united against an innocent victim. But archaic people actually believe that the victim is guilty. They persuade each other of that fact. And there would be no scapegoat phenomenon if there were no conviction of a guilty victim. They will invent that person if he does not exist already. Take the patricide and incest of Oedipus. The most essential thing about the myth is that Oedipus is supposed to be guilty. He himself doesn't know about it, but he's guilty. Freud believed that parricide and incest were unique to the Oedipus myth. He didn't know mythology well enough. He has incredible insights at times, but in this case, he was totally wrong. Parricide and incest are everywhere in mythology. When a crowd gets in serious trouble, it always manages to find patricide and incest, or some other taboo that has been broken, and they kill in the name of that taboo. But it's obvious that it's a false reason. It's the type of reason you get when everybody agrees mimetically. Everyone points the finger and shouts, "He did it!" But what did he do that was so wrong? Someone will say parricide, someone else will say incest, and everybody will pick it up and believe it for no reason at all. That's the absolutely random part of the myth that Freud incorrectly saw as absolutely determined by the nature of humanity.

S.G.: We know today that minorities are often most susceptible to scapegoating.

R.G.: This is true, although there were probably no ethnic minorities in most archaic communities, which in the beginning must have been fairly isolated communities—but there may have been in larger societies as a rule. But you find something in myth which is a proof of scapegoating and of what we would consider the minority status of an individual. We know that mythical heroes, or witches in the Middle Ages, very often have physical defects. And if you look at Greek myth, Oedipus limps. Hephaestus limps too. But some are hunchbacked; some are one-eyed like Odin, the great Germanic god. The number of gods who have a physical defect is enormous.

Why is this? Because a crowd will tend to move against victims that are easy to spot, which have some physical defect or something else that makes them noticeable. And probably in terms of evolutionary theory, it links with privation. You end up picking the animal who doesn't run as well as the other, not necessarily because it's slower, but because it stands out. There is often a visible defect: it's smaller than the others, or it limps. And therefore in a mass of zebras you cannot isolate anything, but if someone is different, one zebra, the lion will rush to this one. And nine times out of ten it will be a good choice because there is a physical defect.

And if you look not only at Greek heroes, but also primitive heroes all over the place, the thing that is amazing about myth is that they have features that are universal worldwide. You see the hero that limps all over the world. So we cannot say that it's something ethnic, something cultural and so forth; it's everywhere. And it makes me feel that it's an objective phenomenon. I'm a realist, you know. I think that texts talk about reality, about real events, and if the texts tell you that so many people limped, there must be a reason.

MYTH

S.G.: You've mentioned mythology several times now, which I understand to be the third part of archaic religion, along with ritual and prohibitions. According to your theory, mythology is a later development than ritual. Ritual is purely mimetic, whereas myth involves a greater level of representational development and ability.

R.G.: At some point in the evolution of cognition and culture, the community needs, and comes up with, an intellectual explanation of what's going on. Of course they do not perceive their own irrationality any more than they can comprehend the real source of their own collective violence. Myths are such intellectual operations. They can be reemployed when conflicts arise.

The people who say that myth has something to do with mysteries of the universe are thinking of Greek myths, which often come to us through the philosophers and are contaminated with philosophical thinking. Philosophy is fine and very interesting, but in this case it has little to do with archaic religion. Archaic myth does only one thing: it tells you how a religion was born—the whole community goes haywire, people get into a circle of violence, and then suddenly they are reconciled. The sacrificial covenant and prohibitions are born from this crisis and its resolution.

Myths try to disguise communal violence behind natural disasters. One very common theme at the beginning of myth is the plague. Archaic societies don't distinguish the plague as a disease. This distinction happened only in the sixteenth century. The plague is primarily violence—an epidemic of people killing each other. The plague or the flood is sacrifice that has gone wild. There's a monster loose in the community and it wants more and more victims. As it gets more and more victims, sacrifice disappears as a source of peace. It becomes violence itself. It joins up with the violence of the community that it normally counteracts. The result is a crisis that seems to be impossible to cure. The more you try, the more violence you create. In myth, after this crisis is described, the focus shifts from the community to a single character. This single character is ultimately accused of being responsible for the whole crisis.

The description in *Oedipus the King* says it all. I always mention *Oedipus* because it's a great play and therefore the foundational myth people know best. Oedipus was looking for a solution to the plague. The people decide he is the problem and not the solution because he's the king. Today that's why there are so many interpretations of the Oedipus tragedy in political terms. No doubt the people are tired of their king, just as the people in Job's village in the Bible are tired of Job. But Oedipus rules over an archaic community, and this community finds a reason for its discontent. The mob he rules over is mad at him and has decided he is responsible for the plague epidemic: he has brought the violence to the community and must be killed or cast out. Oedipus obligingly blinds himself by gouging out his own eyes and is then expelled from the community. In the Oedipus myth there is a kind of trial. This is highly civilized by comparison to archaic myth, in which the crisis at the beginning ends up with an actual lynching.

In Australian myth, for instance, it's a whole community that rushes against the culprit and kills him. There are references to animals capable of charging collectively. The Blackfoot Indians have a lynching myth in which buffalos lynch either a man or another buffalo. But in Australia, which couldn't be influenced by the Blackfoot Indians, it's the kangaroo that kills another kangaroo, which becomes the kangaroo god. And all over Africa, too, you have lynching myths.

But scholars don't want to talk about these lynching myths or to wonder why there are so many. We even find lynching myths in ancient Greece. The most famous mythological cycle, Dionysus, recounts many episodes, and they all end with a lynching. The word "mania" comes from the Greeks and the Dionysus cult—it means homicidal mania. It refers to the climactic moment in the myth, the moment of sacrifice. A small animal is chosen so that the faithful will be able to tear it apart and eat it alive. Have you seen a classical scholar wonder why it's always a lynching? That would be a decent question, but one is not supposed to ask questions about violence. It is pushed under the rug.

S.G.: And yet we don't have myths exactly like these anymore.

R.G.: Today, we would still have myth if we were not the society we are. The Middle Ages still had half-formed myths surrounding what we call the "witch hunt." What is a witch hunt? The community is in crisis, or anticipates one, and begins to look for a culprit. And you usually find a lonely widow, or an isolated person, and you accuse them of all sorts of crimes. It's a scapegoat phenomenon. The Oedipus myth is nothing else. If we started to compare witch hunts in countries where they remain common with mythology, we would see that it's exactly the same thing, except that in mythology the event crystallizes into a coherent sacrificial cult.

S.G.: Why has the witch hunt as such disappeared in the West?

R.G.: The amazing thing about the witch-hunt epidemics of the Middle Ages is not that they happened, but that they were the last ones. We are the only society in the world that has done away completely with that sort of thing. That's why it has become unthinkable to us. But the Middle Ages are very important for this, not because they are bad, not because Christianity is bad, but because it was a period where the influence of the Bible was becoming so important that this sort of thing was becoming more and more impossible. You're going to tell me that it still happens. It's true. But we understand it. We say that it's scapegoating. We never believe it's the epiphany of a god.

THE JUDEO-CHRISTIAN TRADITION

S.G.: The last major development of your Mimetic Theory is your discovery that the Bible treats these phenomena differently than myth, and that the influence of the Bible, in fact, is the primary source of our modern awareness of scapegoating.[5]

R.G.: Yes. In the Gospels and in the Bible you have the same events as in mythology, but they are seen from the point of view of the victim, or the people who join with the victim. I will give you a few examples. Joseph is not a king like Oedipus, he's a brother. But unfortunately for him he has eleven brothers. And they are all jealous of him. At first they wanted to kill him. They didn't. Instead they dip his coat into blood to show the father a wild animal has devoured him and then they sell him into slavery. They scapegoat poor Joseph. And this is so true that Joseph comes close to enacting revenge, and he ultimately puts them through a scapegoat trial before finally forgiving them all.

Now look at the Psalms. Some people tell you the Psalms are vile and you have a victim that curses his fellow citizens. But what is the situation? In probably half of the Psalms the narrator is surrounded by a crowd. The crowd is moving in to close the circle, becoming more and more a threat to him. The crowd is compared to bulls, to dogs, or to other lynching animals. The narrator is the mythical hero on the point of being lynched, complaining about the lynching which is going to take place, and calling on God to stop it.

Or take Job. Job presents himself in the dialogues as the darling of his community. He was in charge, completely. He was a kind of dictator. And suddenly the entire community turned against him. Now they hate him and they find him guilty of all sorts of things. And they delegate the three friends, who are not friends at all, in order to force him to confess as if he were a Soviet leader in the 1930s. That's the difference between Job and Oedipus: Job protests until the end.

The prophets are another example. They tell the people that if they continue in their ways, they're going to disappear. They are very unpopular, and they are terribly mistreated. The most autobiographical of the prophets is Jeremiah, who shows himself as a "scapegoat" for his community. Finally, there are the two incredible chapters of Second Isaiah (52, 53), which describe the lynching of the suffering servant. The texts tell us that the suffering servant is the sort of man that the people don't like, that crowds don't like, that people want to destroy.

The Gospels are also centered on this type of phenomenon. The Lamb of God is a much nicer expression for the scapegoat. The Lamb is a sacrificial victim. Jesus is killed for reasons that have nothing to do with him. He says, "They hated me without a cause," meaning he was an innocent victim, a kind of scapegoat.

The anthropologists at the end of the nineteenth and early twentieth century saw this similarity between the Bible and myth to some extent. They felt that Darwin was marvelous and that his theory dealt a great blow against religion. So they set out to explain the origin of religion by showing that it was only a bunch of myths, that the Bible was only a bunch of myths. They were all racing to write the final theory of religion, to give religion the final blow after Darwin. I think they came very close to the truth. They were right in the sense that they saw the structural similarity between the Bible and myth, the presence of a big crisis moving towards a single victim, who is seen as responsible for the whole thing, and who becomes the savior of the community. But the difference is so plain, so evident, that they looked right past it. Religious people are as guilty as the scientists in this respect. The scientists are naive in thinking they are going to reduce the Bible to a bunch of myths, and the Christians are foolish not to trust the Bible enough, not to be scientific themselves, and say: Let's push the comparison as far as we can and see what there is behind it. If you do that, you discover the truth that all mythical heroes are believed to be guilty from the point of view of the event, as described by the mob—that is to say, from the point of view of the resolution, the return to peace.

But the Bible tells us that these same victims are innocent, beginning with the first murder. Cain and Abel are not the same as Romulus and Remus. The Bible is for the victim right away and tells you the murder is wrong. "What did you do with your brother?" No myth will ever tell you that. It's absolutely essential. It makes all the difference in the world. All our civil rights come from this.

Joseph is shown to be innocent. The Bible tells you that his brothers are jealous. Job was innocent. The suffering servant in Isaiah is innocent. Jesus is innocent. In fact, the Gospels themselves show you how scapegoating works more completely than any other text. The marvelous thing about them is that they show you the creation of mythology, and the fact that the death of Jesus is a crowd phenomenon.

S.G.: Can you give an example of this?

R.G.: The most important theoretical text in my view is Peter's denial. The psychological interpretation of Peter is unfortunate: they say Peter is a weak individual. No. He's the most typical man there. When you get into a crowd that is entirely mobilized against an individual, you join that crowd. Very few people can withstand it. Peter's denial shows you that this is the principal of unity against Jesus. It means nothing individually about Peter. It's foolish to say: If I had been in Peter's place, I would have resisted better. No. Peter was the best of the disciples, and as soon as he gets into a crowd that believes Jesus is guilty, for a few seconds, he joins the crowd. We are crowd joiners because we are terribly mimetic. It is much easier to join the crowd than risk being killed ourselves. That's why human culture begins in scapegoating, and why we had sacrificial religions. But the Gospels show you Peter in order to teach you differently. The error of the Christian Church is to worship the cross, which is, in a way, idolatry. What must be worshiped is the awareness of what Jesus is doing.

So the Gospels are built exactly like a myth, with one exception. There are a few people who eventually secede from the crowd to say, "It's not true. Stop everything. The victim is innocent." This never happens in a myth. There are a few sentences in *Oedipus the King* that show you that Sophocles had doubts about his own myth; but if he had changed the myth, the crowd would have lynched him right there at the theater in front of the whole population. The theater fulfilled a religious function, and so Oedipus remains guilty for the crimes attributed to him by his community. Not so for Jesus. The Gospels in the Bible are totally different.

The Christian religion is a constant struggle against scapegoating, but Christians never talk about the incredible structural closeness between myth and the Christian story. Both myth and the Gospels are fundamentally scapegoat stories. The difference is that in myth you are always on the side of the mob without knowing it, and in the Gospels you are for the innocent victim. Many realize this concretely by not being on the side of the mob, but to formulate the difference conceptually is another matter. All anti-Christian arguments, supposedly "scientific" arguments, show you that the Gospels have the same structure and content as mythology and therefore proclaim that Christianity is yet another myth. But there is one difference, which is all the difference in the world; in the Gospels you are not against Christ or for the persecutors, as you would be in a myth.

S.G.: The crucifixion of Jesus, and the Bible in general are usually criticized for being too violent, and you're saying, in a way, that the depiction of violence in the Bible is its strength, because it tells the truth of the predicament of human nature more accurately.

R.G.: I remember a French writer from the '30s saying something to the effect that "In the Bible, there is that foul smell of scapegoating, which is not there in mythology." It is thought that the Bible is full of scapegoating and violence while mythology is clean, full of nice Greek columns, and so forth, but the opposite is true. To have a scapegoat is to not know that you have one. You wash your hands clean of the violence and write a very nice story. However, as soon as you realize you are scapegoating, it doesn't work, and you repent of what really took place. The scholars today tell you that the Gospels are a bunch of crooked texts that try to hide what really happened. But if they were a bunch of crooked texts, the first thing they would have done away with is Peter's denial, since Peter was the head of the Church. They didn't. Peter's denial tells you the essential thing about humanity. Most of the time people fight each other, but when a problem arises they will all join the crowd against a third party, and that is how we have managed to be a society. But the Gospels and the Bible tell us that the time has now come to do things differently.

Part 2

Mimetic Theory and the New Science of Imitation

S.G.: I would like to turn now to the current research on imitation in the experimental sciences. You have spent your entire career writing about the importance of imitation, and yet, as you mentioned before, many have viewed imitation as irrelevant. It was never really taken seriously throughout most of the twentieth century. Now, however, as you approach the end of your career, imitation is suddenly a very popular topic—revolutionary, in fact, with respect to the cognitive and social sciences. What is your impression of this development?

R.G.: The focus on imitation is very positive. I think it would be impossible to do away with imitation now, as a result of the recent work that has been done. We tried to do that in a way already. The social sciences have been predicated on the absence of imitation, which is really not justified. It's actually quite ridiculous, because our culture talks about imitation all the time, only in a very negative way.

S.G.: You mean that imitation is stigmatized?

R.G.: Yes, we are obsessed with not imitating. During the whole twentieth century, imitation was out of fashion because it was despised as a behavior. You had to be original, singular. The world was becoming more uniform all around us, and the intellectual reaction was to deny this at the individual level. Everybody was for originality. It was already important in the nineteenth century, but in the twentieth century you had to be original at all costs. All the antics of the surrealists were about how to be original in a world where everything has been done before. Let's do anything—ridiculous, obscene, whatever—provided it is original. Culture changes in many ways as a result, but not always in a way that represents real progress.

The world of fashion is a good example. We know that fashion changes constantly. Where do these changes come from? They have to be imitated; you see some guy with a different type of suit and you want it too. What is fashion? Is it a notion that is purely romantic and subjective, which has no scientific basis? No, it is dependent on imitation, because if you look at the history books, you can turn to any page and date it immediately based on the way people are dressed. I'm fundamentally a historian, and I have found that this is true for just about anything.

Of course advertisers know all about mimetic desire. They are not expected to show you that a given product is really better than another product. Instead, they show you beautiful people playing on the sand, on a beautiful beach, drinking a soft drink. And because you want to be like these people, you want to drink that soft drink too. At some level you think: Maybe if I drink that soft drink I will be more like the people on the beach than I really am, because deep down I don't feel that I am like them at all.

S.G.: Despite its general neglect in the twentieth century, there is now consensus among a great number of researchers that imitation is pervasive in human life, and functions at a level below conscious awareness. In fact, imitation is now considered the "default" human behavior. How do you think your theory would have been received had it been introduced in this more favorable climate? Do you think that a lack of understanding of the function and pervasiveness of imitation at the level you describe contributed to a lack of understanding of your theory as a whole?

R.G.: Sure, without a doubt. What I'm saying to you is already present in my first book, but nobody paid any attention to it.

S.G.: To your analysis of mimetic desire?

R.G.: Yes, and especially to my analysis of mimesis as conflictual.

S.G.: Why do you think the relationship between imitation, desire, and conflict remains so unrecognized, or taboo?

R.G.: It's very threatening and humiliating for us to know that we owe our desire, which is supposed to be mostly ours, to other people; that we are fundamentally dependent on each other in this way. So very often we cover up the fact that we desire the same thing for the same reason under the pretense of anteriority—I wanted it first. It's always the other person who is the one imitating.

One could carry this to political problems and social differences. Each side denounces the other, either for envy or for culpable display of luxury or power. For example, the same people who are against social laws are against the Left and feel that others are acting out of envy. However, one should answer them and say that you are right so far as you yourself didn't motivate that envy because you need it to make sure you're really the superior one. So it would be very important to be able to tell with a certain degree of accuracy who is provoking who. Is the reaction legitimate given the average situation of the same type in the same area at the same period of history? We'll probably never be able to do that. Nonetheless, instead of having a static definition of political and social problems, we need to make them more dynamic and intersubjective, which is rarely done. Conflicts of ideas can be important, but very often they are a cover for a rivalry of desires, and we hide the fact of mutual desire under the pretense of ideological differences.

S.G.: What is your impression of the way imitation researchers have addressed this conflictual dimension of imitation and desire?

R.G.: I'm not a specialist in this area, but my impression is that the empirical sciences have not been interested in that at all. In fact, they seem interested in doing away with the contradictions inherent in imitation, or at least in downplaying them. There is an unconscious avoidance of the problem of human conflict. Our minds tend to chase the problem away by not putting together what should be put together. Mimesis implies a positive attitude toward the object that the other person desires, and therefore toward that other person too, and then suddenly this positive relationship is brutally interrupted by conflict due to competition over the same object. This conflict in turn is misunderstood, and this misunderstanding creates more hostility and ultimately violence. We desire the same thing and we are friends; we desire the same thing and we are enemies. There is no transition between the two. Inevitably this is disturbing for us to realize, but in fact it is the easiest thing to interpret, and it's kind of surprising that no one has interpreted it before. Our ability to share desires conceals conflict at every turn. It is a matter of questioning language. You can use a whole vocabulary pertaining to positive relationships, but that language is inevitably inadequate. If it expresses one thing, it will not express the other simultaneously.

So the main problem is to define this contradiction; the whole twentieth century was against imitation, but before the twentieth century, imitation was extremely popular and thought to explain almost everything. Perhaps you could say that the express purpose of the Mimetic Theory is to link these two different periods, and to bring out the contradictions and the difficulties inherent in them.

S.G.: The contradiction of human imitation is becoming more and more apparent from the perspective of cognitive science. Here is a passage from the introduction to the most recent and comprehensive volume on imitation research:

> How should we respond to the irony of imitation: that the capacity for imitation appears to be a distinctive feature of human nature and may well be part of the basis for other distinctive features of human nature, such as mind reading and language, which together set us apart from other animals? Yet at the same time our innate, automatic tendencies to imitate can also threaten our conception of ourselves as autonomous and deliberative in ways that no other animals are.[6]

R.G.: So there you have the contradiction of imitation in humans and its potential threat. However, notice that the drama and conflict that is born out of this contradiction is not discussed. For me, this drama is absolutely essential. We have the contradiction of the goodness of humanity on the one hand, and on the other, we have its immense violence. What does this mean? Should we speak of evil? No. I think we have to use scientific language. This contradiction means that you cannot stop imitating the violence of your opponent any more than you can stop imitating the kindness. Kindness escalates and turns into what we call love, which animals don't exactly have. But it escalates the other way too, and it turns into deadly violence, which animals don't have either. There you have two characteristics of humans, which can be founded empirically in imitation.

S.G.: In your theory, the emphasis is on mimetic rivalry, but what you have just said suggests there are also very positive aspects of mimesis.

R.G.: Well, that mimesis itself could be good or bad I think is meaningless. It all depends on what is at stake in the phenomena. I don't think it would serve any serious goal to divide mimesis into two. It would be a pseudo-solution.

S.G.: Your literary discoveries reflect an unprecedented sensitivity to the way human relationships work, especially to the presence and nuanced nature of imitation. Can you speak about the difference in prestige or attention that is given to discoveries that come about as the result of experimental science as opposed to theoretical insights based on literary research?

R.G.: This is a process that started with the Enlightenment, and which is on the way out today. Everybody knows that you don't discover Einstein's theory through small methodological steps taken one at a time. No, such insights come all of a sudden, and then you spend your entire life interpreting them. In a way, the need for insight is felt again, but at the same time it hasn't displaced the need for science. The problem of readjustment is so great that I don't think it is being tackled very efficiently. Now it is fashionable to say that unifying systems and methods are impossible. If you are systematic, you are necessarily wrong. But every great scientific discovery is approached systematically! Only systematic thinking works, in a way. Maybe it's impossible to a certain degree, but we are always aiming towards it. And my theory is a systematic theory.

S.G.: According to Jean-Pierre Dupuy, there is potentially a positive relationship between experimental science and the Mimetic Theory, but it should not be considered a hierarchical relationship in which one has to be founded on, or proven by, the other.[7] Can you comment on this?

R.G.: Fundamentally, I think Dupuy is right. All scientists know that many scientific innovations consist in importing into a neighboring area something which has been invented elsewhere, something which has not only worked and produced things, but suddenly illuminated a problem that until then was totally obscure. Much of discovery involves transporting something from one area to another. It's very tricky, but if it works, it works.

If science could find a way to translate imitation into foolproof equations and operations, I would be all for that. I think quantifying these matters is very important. And indeed there is something very scientific in what we have been doing. But ultimately, I'm more interested in insights than organized research—overall insights that demand extensive commentary and interpretation and that are not the result of experiments. I'm afraid if it's not guided by a moral purpose of some kind, with application to real relationships, science will only produce empty formulas.

IMITATION IN INFANCY

S.G.: One of the main findings that sparked recent interest in imitation by experimental scientists is the discovery that infants are able to imitate immediately at birth.[8] As a result of work done by developmental psychologists, such as Andrew Meltzoff, the intersubjective capacity for imitation is now considered the starting state of human cognition and sociality. How does this research impress you?

R.G.: That imitation was found at birth is very important. This is similar to my own view of imitation that I have described in my books from the beginning, that it is something preverbal and pre-representational. I wouldn't even have imagined that people didn't see this role of imitation in childhood. But apparently they didn't.

S.G.: Before Meltzoff's research, the common understanding was that children learned how to imitate later on in development as a result of gaining certain representational skills.

R.G.: Sure, there was the idea that imitation was learned. Would you agree with me if I said that the general trend of these recent discoveries is that there is much more in us that is innate and comes earlier than previously believed?

S.G.: I think the answer to that question, according to Meltzoff's developmental research, is both yes and no. In one sense there is more built in, especially where social reciprocity and understanding of others is concerned, than was stated by traditional theories. On the other hand, that bit of innate equipment allows for a more powerful type of imitative learning that takes humans beyond the initial state. Much more is learned early on, and at the same time, there is much more innate equipment. Both are considered together. Piaget didn't think we had that kind of early learning, especially in terms of social relations. For both Piaget and Freud, at this earliest stage of development, there's barely a self, let alone an "other." There's no real interaction between the two. So our views have changed quite a bit as a result of this research.

In fact, these early imitative exchanges have also been linked to fundamental motives, whereby the ability to share psychological states is a qualitatively distinct human ability that is desired in and of itself—for no other reason than to have someone else share in our same experience. Ann Kruger notes, "The urgency with which children request another's attention to the action they perform suggests that the action itself is not enough."[9]

R.G.: This intentional sharing of a common focus for its own sake is a definition of the human way of communicating experience with someone else, which certainly animals don't do. I agree with that completely. We try to get people to admire us. This is also the beginning of snobbery, in a way.

S.G.: This early form of desire for social recognition is seen as an innocent thing that children do, but you're saying it is also the basis of snobbery. In a number of contexts the innocence of children and humans is presumed, but you don't presume that innocence.

R.G.: No, one shouldn't, because we know that these things very quickly turn into rivalry, even early in childhood. If this were not so, we would not need culture. We would not need parents to intervene.

S.G.: You would agree, however, that there is a cooperative mimetic that is originally there as well, which is the essential self-giving from a mother or parent to their child. To have a rich cultural and relational understanding, it seems to me, you need this other form of imitation.

R.G.: You're right. In a way, the positive relationship between mother and child is much more important to culture than all the fighting. But in the Mimetic Theory, you're looking for trouble constantly, because once initiated, mimetic rivalry and violence become all-consuming. So you need a way to understand how human culture was created and survived in the first place under such conditions. Scapegoating violence solved this problem and allowed societies to structure themselves in a way that made room for other forms of mimesis. So it doesn't mean there is only violence. You are quite right to make that observation.

S.G.: At the same time, just to argue your point further, Meltzoff notes that while infant interactions can be regarded as reciprocity in the positive sense, very soon there is sibling rivalry and rivalry for the parent's attention.[10] It seems to me that this involves imitation as well, because when the child sees the mother with the newborn, the child feels rivalry for the newborn and then imitates the baby in order to get the mother's attention. So early on, children are seen acting like, even mimicking, a younger baby. That seems to be a form of rivalry.

Even in the close relationship between the mother and infant, which we like to think optimistically as entirely positive, there's also a tendency for the baby to pull in the mother in such a way that she can feel exhausted and resentful about spending all of her time with the child. When you look at what the mother does, the child is imitating the mother, but there is also a tendency for the mother to act babyish. The phenomenological feeling that parents describe after taking care of a child is that they enter the child's world and kind of give themselves up, which is really another way of describing the process of undifferentiation. As a result, there's a pushing away that sometimes erupts in violence against the child in order to reestablish oneself.

The opposite is also true as well, in that the baby will often need to pull away and differentiate from the parent. Often times the parent blocks this attempt by continuing to intrude on the infant's experience. In fact, using split-screen video analyses, Beatrice Beebe has demonstrated how mothers and their infants (only three to four months old) can participate in highly synchronized and imitative behaviors that are mutually escalating, both in the positive sense of being enjoyable as well as generating intense negative experiences.[11] For example, Beebe notes a "chase and dodge" interaction where the mother increases her pursuit of the infant's gaze while at the same time the infant is trying more and more to turn away from the mother. The more the mother chases the infant for further interaction the more the infant dodges her, resulting in a form of what Beebe calls "mutually escalating overarousal." In one case, the interaction "bursts out of control" and comes to a halt after the child is sobbing and finally throws up. So even in the early mother-infant relationship, there is also some violence, rivalry, and negativity.[12]

MIRROR NEURONS

S.G.: Part of the problem with earlier conceptualizations of imitation is that there was no way to conceive of such intersubjective relations operating early in life, since imitation was thought to be based on the later development of formal representational thinking, or associative learning. This is where the discovery of mirror neurons has been important for corroborating theories of immediate imitation based on behavioral observation that suggest there must be some type of neurocognitive wiring allowing for this early level of cross modal matching, or shared psychological states. Do you view the discovery of mirror neurons as important? Is this something that interests you?

R.G.: Yes, it interests me, but again I can't say that I am a specialist in this area. The fact that mirror neurons respond to the same thing whether one is observing or participating I think is very important to my ideas. There are people who have tried to discount that for various reasons, but I believe it's inaccurate to say that it has nothing to do with my actual work. But in a way, we know that what mirror neurons demonstrate is true, even if we don't have a precise explanation for it. For instance, if you watch a climber on a mountain, or even a movie of a climber who is about to fall, you yourself will automatically be in the same position and then do all the gestures to save him. It is a natural thing to do. That's what the mirror neurons are about, as I understand them.

S.G.: What is important about the discovery is that it supports a shift in theory from a mental process of abstract representation to mental processes that are subpersonal and automatically simulating current actions and feelings of others. The standard way of thinking about our ability to empathize was through more conscious reflection, which is a more cognitivist perspective. Even though we can think more consciously and intentionally about others, the more recent understanding is that empathy begins at a much lower level of embodied activity.

R.G.: In other words, there was previously a problematic of representation, which you can now do away with completely.

S.G.: But in the doing away . . .

R.G.: You keep something . . .

S.G.: You provide a richer context to think about the mental power of imitation, or mimesis. Your actions are simulated in me at an unconscious level while I watch you, meaning I'm more likely to do what you just did, whether I'm aware of it or not. I can certainly think about and reflect on your thoughts and feelings more intentionally at some point, but below and before that, there is an intersubjective field in which we are influenced by the embodied experience of the other.

Comparative Psychology and the Origins of Culture and Religion

S.G.: The work on mirror neurons also touches on another important area of research, which is comparative psychology; this includes understanding the differences in imitative abilities between humans and nonhuman primates. Monkeys have mirror neurons, but they do not imitate like humans, so there must be a way to account for the cognitive and social differences between species and their role in the origin of human culture. Do you believe this type of research is important for the further development of your theory?

R.G.: I believe there is specificity of the human, and that you must not shift from the nonhuman to the human without great precautions. Clarifying these stages of development and evolution is very important.

It's interesting that monkeys have a reputation for imitation, which is totally undeserved. Monkeys are so close to humans that we take for granted that they should imitate, when really they imitate very little, whereas it's a very telltale behavior for humans.

S.G.: The research is very clear that monkeys, if they do imitate, do not do so anywhere near the level of which humans are capable.

R.G.: Aristotle was right in saying that humans are the most mimetic of all animals, because we know very well that the great apes are close to us, and they are more mimetic than other animals. They are mimetic enough that we laugh at them, just as we laugh at ourselves, but we are obviously much more mimetic than they are.

S.G.: Many contemporary theories of cultural origins, such as those proposed by Michael Tomasello and Merlin Donald, give imitation, or mimesis, an essential and defining role in the transition from nonhuman primates to humans.[13] However, these theories do not mention at all the role or presence of conflict and violence as a key variable that has any sort of impact on these evolutionary processes. Your theory, on the other hand, recognizes the problem of violence from the very beginning as an essential variable in understanding how and why human culture came into being. Not only that, but it provides a type of continuity between our origins and the dominant role that violence still plays in the world today. Can you elaborate on your impression of these more recent theories and why you insist that humanity necessarily begins with a violent crisis?

R.G.: Well, I think that the Mimetic Theory has to be better inscribed in evolutionary theory. I'm a Darwinist. I believe in natural selection. And one of the contemporary theorists, Richard Dawkins, said we have to explain culture too by means of evolutionary theory.[14] And he invented what he calls *memes*, which are units of culture. Immediately after that, people picked up on it, and now there are many evolutionists or theoreticians today who discuss culture in terms of *memetics*, but they haven't learned that *mimesis* is potentially conflictual. It's obviously the greatest source of conflict in human life.

All communities have crises, and I believe the mimetic crisis of the type I have described is a special human feature. Humans are an animal of crisis. The real problem with our social sciences is that they've never learned that. The social sciences confuse the science of humanity with what human beings say about the order of their community. If you're a sociologist, you study that order according to what people say about their own culture. If you're a psychologist, you talk about the self, according to what the self is telling you. But the question is: when a society gets into a serious crisis, who is in charge? If you listen to the government, of course you're listening to their propaganda, but very often the government is no longer in charge. None of the standard rules apply. What is a crisis? The social sciences have not been able to define it. There is a great political scientist who first said that. His name is Carl Schmitt. He said a social science should first be the science of crisis. It's very easy to define who the government is, but in most societies the government is not the most important authority. You have to study a crisis in order to see who is in charge there. Who is fighting whom? What's really going on? Now in a way, the Mimetic Theory would like to do that, because humanity necessarily starts with a crisis. It's true that it's not enough to talk about biological changes. You have to talk about cultural changes, because we know that the last stages of the evolution of humanity are both cultural and physical.

For instance, the human infant needs so much care for so many years, that it is impossible to think humanity could have developed without any culture—in other words, without any taboos on violence—on violence inside the family. We know that the male would not tolerate being separated from the female for as long as is necessary in a human family for the early development of young children. Therefore there must be a mix between the two. We have to find mechanisms that are part of the culture and which at the same time should be quite mechanical, that is, happen spontaneously, because no human was capable of inventing them on their own. And I think that religion goes back that far. The incredible power that religion has over us must be there already at the beginning of humanity.

So we have to be able to talk about religion in a scientific way as well, which will be neither anti-religious as it is today, nor pro-religious in a spiritualist way. The discussions between the creationists and evolutionists, and even between spiritualists and materialists are becoming outmoded. The question is to find ways to talk about what happens when dominance patterns disappear and when culture, a human type of culture, is beginning to take over. And how can we know about that? We have to look at the structure of religion: at the nature and function of religious rituals and prohibition.

S.G.: With respect to escalating levels of violence in the evolution of primates, chimpanzees have been found to collectively kill their own kind. Large groups of male chimpanzees, for example, have been observed going out together, patrolling the border of their territory, and finding other chimpanzees that they then brutally assault and often beat to death.[15] Does that finding contradict your theory in terms of the uniqueness of human violence, or does it help fill in the gap from animal violence to human sacrifice?

R.G.: I think that it helps bridge the gap. I don't insist on the uniqueness of violence in humans. It's all right if the facts show otherwise. They don't show it for all animals, but they do show it for chimpanzees, who are quite special in the animal world for their proximity to humans.

S.G.: In your view, the transition to human culture and religion came about as a result of changes at the *inter-dividual level*, or as a social solution to this escalating problem of rivalry and violence in primates and early proto-humans. Contemporary perspectives from the field of cognitive science of religion, on the other hand, depict the origins of human culture and religion as a result of changes at the *individual level*.[16] Such theories stress that what is unique about the evolution of the human mind and culture are changes that take place in a particular brain structure or system. This mode of thought presumes that the source of anything that is important about humankind, such as religion, is an outcome of these "inner" or individual changes in cognitive capacities.

According to your theory, however, the origin of religion and culture is an entirely interpersonal process, and the scapegoat solution occurred at a time in human evolution that was prior to many of the cognitive characteristics that are considered important to human cognition. The scapegoat solution to the problem of ubiquitous violence must be repeated often to maintain the peace. Consequently, the rites and prohibitions that constitute religion grow up to institutionalize the scapegoat mechanism. This is why religions worldwide have certain core similarities. In other words, for you, religion is not a byproduct of the evolution of mental tools, but rather religion is an environmental force that causes further social and biological evolution to take a particular course.

R.G.: Correct. My tendency is to see scapegoating violence, particularly the kind that must have involved the entire community, as creative of human culture. I'm not an expert in the actual cognitive science of human origins. I read avidly what I find, but it doesn't go beyond that. And my work has brought me to the conclusion that mimesis and violence are essential to account for human origins. So it is remarkable to me that these contemporary theories do not take this into account.

What is amazing is when there is no violence in a particular culture, not the fact that there is violence. Conflict should be permanent, constantly there, since we imitate each other and are prone to mimetic rivalry. What is interesting is to explain how a society manages those conflicts. If a scapegoat phenomenon puts an end to these hostilities, it will be repeated artificially. This is the definition of sacrifice, which for me is the beginning of human culture.

S.G.: But how is rational reflective thought—what we consider a defining characteristic of our species—born out of such irrational activity; out of the type of social chaos and causal misunderstanding that is characteristic of mob behavior and scapegoating violence? In other words, how is scapegoating the foundation that sets us on the trajectory toward our higher cognitive abilities, including language and rational reflective thought?

R.G.: One should say, scapegoat unanimity is the beginning of rationality. It is false rationality, but why is the scapegoat chosen unanimously? Because the community spontaneously discovered and subsequently believed in the majority principle. They have become democrats right there. They say, "If we are all unanimous, how could we be wrong?" They are wrong, but they are more right than other animals that cannot have unanimity. Of course this is not a matter of conscious reflective thought initially, but of mimetic repetition. The self-conscious unanimity of scapegoating is the beginning of humanity. It's the beginning of rationality. It's still irrationality, no doubt, but it's a beginning. You have to go through that phase in order to reach rationality.

S.G.: So the tendency is to ignore violence in archaic culture and to view it as a late-developing phenomenon. We prefer to view the beginning of humanity as fundamentally peaceful and then blame our violence on more contemporary religious or cultural beliefs.

R.G.: We are in a privileged position to be able to talk about the essence of human community as positive—and to be sure there are many positive aspects, but there is always some kind of violence lurking in the background, which we would prefer to keep hidden and not mention in the same breath.

S.G.: I would like to connect this discussion on the presumed innocence of early humans to our previous discussion on the presumed innocence of the early parent-infant relationship. I believe the vast majority of crimes where children, especially younger children and infants, are abused, harmed or murdered occur within the family. And the Jewish story of Abraham and Isaac is a parent attempting to sacrifice their child. A parent is about to murder their child because God asks him to, and he's willing to do so. And that is what leads to the covenant. So in almost the same way that the sacrifice of Jesus has been viewed as the founding of Christianity, the near-sacrifice of Isaac is related to the founding of Judaism. And we know that the ritual of sacrificing the firstborn predates both of these traditions.

R.G.: You're right to bring this up. The sacrifice of the firstborn is found in many places. It's almost a universal custom at some point. And I think the Passover story is a good example.

S.G.: For the ancient Hebrews, once the angel stops Abraham from killing Isaac, a ram appears magically. And that's the beginning of the thousand-year tradition of the sacrifice of animals, and that's the substitute for human sacrifice. But Jesus's sacrifice, that's the end of all sacrifices, including animals. That's the last sacrifice. And then for the rest of the time, you remember it so as not to repeat it.

R.G.: Correct. But the shift to an animal comes before and has tremendous importance from a human point of view, from an evolutionary perspective. The near-sacrifice of Isaac is the only text we have about the shift from the sacrifice of the firstborn to the sacrifice of an animal, which probably, in the evolution of humanity, is an immense thing. I'm surprised that there is no interpretation of this. It's so obvious, the shift from human to animal sacrifice. And the Bible shows you steps of less and less sacrifice, of less and less precious victims being killed. Every step is a descending staircase towards no sacrifice in the archaic sense.

THE REVELATION OF SCAPEGOATING AND THE MODERN CONCERN FOR VICTIMS

R.G.: I think in a way the course of history is a constant revelation of scapegoating, and today this revelation is fruitful in the sense that it shows you really, with many exceptions of course, where the violence lies, where the victim is. It can be misused, distorted, and so on, but there is a historical trend toward the revelation of scapegoating.

S.G.: You believe that this trend is linked to our modern concern for victims.

R.G.: Yes, what is important is the emotional leap in our attitude towards victims. We have a unique attitude. We don't realize that. On the one hand we can say how bad we are, that we are so violent, but at the same time we are the only society in the world which has ever denounced and protested its own violence. You can go back to any society in the past and you will find there is no society that ever worried about its own violence like we do. It was simply regarded as one of the facts of life, about which you cannot do anything. So we are constantly accusing ourselves as if there were better societies before that we could give as examples of nonviolence compared to us. Voltaire, during the Enlightenment, believed that so much, that when he wrote *Candide* he decided to find a counterexample. He must have looked for some place that he could use, and he decided there was none, and so he invented a false one, his El Dorado, which is a myth: the place without violence. Voltaire was the first one to have that attitude toward violence, which we have, which is: everybody is responsible for violence except me, except us. People before talked in terms of evil, but not of some kind of objective violence. What does violence mean objectively? We cannot define it. I don't think we should try. It's always a relationship.

S.G.: Can you give an example of the way in which scapegoating is revealed and reversed in the modern world?

R.G.: I usually give the example of the Dreyfus case. We all know about the Dreyfus case in France at the end of the nineteenth century. It's not very good for the government to have a Dreyfus case. But if you have people who are treated like Dreyfus by the judicial system, it's better to have a Dreyfus case than to not have one, because it means that the judicial error will come to light

Dreyfus was a French officer. He was Jewish—a favorite scapegoat of the culture. And he was accused very specifically of having written a document which included some classified secret and had been sent to the Germans. France was in a crisis. France knew that the war with Germany was not the end of it and that the Germans were stronger. France was worried about itself, and especially about its army. So it was nice to have found a good traitor to comfort people. The majority of the nation believed in the guilt of Dreyfus, who was then sent to do forced labor in French Guyana. But after a while, some people began to have doubts. These people were not revolutionaries. One of them was the vice president of the Senate, a very conservative man and a very well-known politician. Another one was a colonel who knew Dreyfus. Indeed, Colonel Picard, who was the first to denounce the situation, had to go to jail.

What does the Dreyfus case mean? Who are the scapegoaters? They are the people who believed in Dreyfus's guilt. And they really did believe that he was guilty. Therefore they would not call Dreyfus a scapegoat. They would say he was a justly condemned spy. The people who started to use the word "scapegoat" were the few people who rebelled against the situation. The word has been in the language since the seventeenth century. A friend of mine, a Japanese anthropologist, said that when you want to talk about scapegoating in Japanese, you have to use the Western words, because Japanese doesn't have them in that demystifying sense. It has many words for the ritual action comparable to scapegoating, but no word that would mean the revelation of the truth behind the ritual. And why do we in the West have it? Because we have the Bible that tells us about innocent victims who suffer from the crowd not taking responsibility for its own actions.

S.G.: Do you believe the Bible is the only text that reveals scapegoating?

R.G.: I don't want to privilege the Bible exclusively. I have given many other examples from other traditions. For instance, if you take the evolution of Indian literature, the Indians have the most incredible scriptural texts. Everything I say about sacrifice they also say. They don't have myth only, they have commentaries and explanations of myth, which are still within the orbit of Vedic thinking, and tell you about rivalry and about the victim as a solution to violence. They tell you everything. And when you get to the period of the great Jewish prophets, you have the Vedanta, and these great texts in the Upanishad. There are so many of them that we haven't translated them all. In the Upanishad you find an empty sacrificial thrust where sacrifice is regarded as murder, and rejected. Later Hinduism went back to sacrifice. But you have a period that is analogous to that of the Jewish prophets. They don't give up the word "sacrifice," but there's a sacrifice which should be purely interior. Very often you have funny stories about the demons being behind sacrifice, the demons exploiting man, and the gods trying to stop it. This anti-sacrificial trend is present in Buddhism in the sense that Buddhism comes entirely from the Indian tradition but separated itself, whereas the texts in the Upanishad are in continuity with the great Indian tradition. We know that Buddhism is against sacrifice, and at the same time, one has to recognize that Buddhism, or the territory where it predominates, has not eliminated earlier forms of sacrifice to the same extent that Christianity has. Neither has Islam. Islam remains compatible with certain forms of animal sacrifice. Christianity, no—and in Judaism, of course, the sacrifices ended with the destruction of the Second Temple.

S.G.: So in your view, Christianity has had a tremendous influence on modernity.

R.G.: It's the simplest truth in the world. But we have not been able to reach it conceptually. There was one thinker I believe who did, and that was Nietzsche, but he became so perverse that he said it was better to be with the violent than with the peaceful. He said that Christianity was against sacrifice, against the necessary victims that a society should make in order to be healthy. That's the whole theory of the extreme Right today. Nietzsche embraced this violence. He made it his own, and he went mad. But Nietzsche, in a way, was better than all nineteenth-century theologians because he understood this difference: he said that the death of Dionysus justified that the victim was guilty, whereas the death of Christ was unjustified, and this knowledge therefore destroys the peace of humanity. And it's true; it destroys the peace that comes from scapegoating violence. Many people, when I say what I'm telling you now, say, "So what?" So what? Everything is at stake now.

The Future

S.G.: On that note, in your most recent book, *Achever Clausewitz*,[17] you turn from comparative anthropology and archaic religion to an analysis of the Prussian military strategist Carl von Clausewitz's book *On War* and its implications for understanding the accelerated course of human warfare over the last two hundred years, including its implications for the potentially devastating future of humanity. What other areas of research or application are you currently interested in?

R.G.: As I get older, I'm more and more interested in history. And one of the things I'm interested in is the situation concerning our current environmental problems. The environment is at the forefront of the world's attention in an unprecedented manner. People are more and more aware of the environmental problems, but in a way they are not aware of the historical change that they signify. It's an absolutely essential event that should not have been unforeseen. It was obvious that technology, and the influence of mankind upon its surroundings, was going to pose urgent problems.

S.G.: Because we now have the ability to manipulate nature in such a way?

R.G.: Yes, and we've been manipulating nature until now without worrying about the consequences. Now suddenly these consequences have become more and more apparent, and therefore we know we are at the beginning of a new era. These problems will have to be integrated into politics, philosophy, history, and all aspects of human knowledge.

S.G.: How can intellectuals respond to the environmental crisis? What disciplines are equipped to meet this new challenge?

R.G.: It demands a theory of human action, and of the consequences of human action. No single science, as science is organized today, can speak about it. It's not enough to talk only about environmental problems, because human responsibility is not taken into account to the degree that it should be. On the other hand, history has never bothered with nature in the past. Past history is political history, which is a very narrow view of human action in the world. We need a sociology that is more dynamic—one that can take into account the changes in the world as a result of the so-called "human project."

S.G.: Would you say that our environmental problems are linked to the mimetic nature of our desires in that we want more than we could possibly ever have or need?

R.G.: Indeed, mimetic desire has played an important role in leading us to the current situation, including the way we have not taken into account the potential effects on the environment in attempting to fulfill all of our desires. Our lack of foresight was inevitable to some extent, but now it is becoming more visible every day. Immense measures for the environment are needed now, which are nothing at all compared to what probably will be needed in the near future. The national system cannot work anymore. At the same time, international associations like the United Nations have never been able to get sufficient power to enforce rules on sovereign countries. This problem is going to become so acute that there should be a change in the attitude of the nations themselves.

S.G.: Are you hopeful?

R.G.: I'm not sure we are ready. Unfortunately, there will have to be major incidents, major problems, before people are motivated enough to change. The melting of the poles is probably the first of such events, but not quite. If the sea rises to a significant degree, maybe there will be a change.

S.G.: Many movies nowadays depict these global catastrophes, such as New York under water …

R.G.: They're not just a joke or science fiction. In a way, they say more about what's on people's minds right now than the United Nations meetings. Sooner or later, there will have to be democratic means taken with the consent of the population, but which will demand global involvement and therefore social and financial resources. This will pose serious problems since there are people who want less and less government and believe that you can go back to conditions in which individuals are completely free and have no social obligation to intervene collectively.

S.G.: You are saying we will have to do away with our Enlightenment view that progress in knowledge and science necessarily translates into social progress.

R.G.: We always want absolute peace, plus money, computers, and everything else, but it doesn't seem to be working quite right at this present moment. Humanity is in danger, and our situation requires a change in all the human sciences—a radical change. It is obvious now that we have to take into account the probabilities and possibilities of our actions, the effects of which cannot be measured or predicted with absolute precision. It is clear that people are beginning to have second thoughts about Enlightenment optimism and the romantic view of human freedom and desire, and all I want to do is increase, a little bit, these doubts.

ACKNOWLEDGMENTS

Scott Garrels would like to thank Trevor Merrill for his help with transcribing and editing this interview and Mark Anspach for his help with reviewing the completed manuscript.

NOTES

1. René Girard, *Deceit, Desire, and the Novel: Self and Other in Literary Structure*, trans. Yvonne Freccero (Baltimore: Johns Hopkins University Press, 1965).

2. Gabriel Tarde, *The Laws of Imitation*, trans. Elsie Clews Parsons (New York: Henry Holt and Co., 1903).

3. René Girard, *Violence and the Sacred*, trans. Patrick Gregory (Baltimore: Johns Hopkins University Press, 1977).

4. See Melvin Konner "Sacred Violence, Mimetic Rivalry, and War," this volume for an overview of the research on violence in non-human primates.

5. See René Girard, *Things Hidden since the Foundation of the World*, trans. Stephen Bann and Michael Metteer (Stanford, CA: Stanford University Press, 1987).

6. Susan Hurley and Nick Chater, eds., *Perspectives on Human Imitation: From Neuroscience to Social Science* (Cambridge, MA: MIT Press, 2005).

7. See Jean-Pierre Dupuy "Naturalizing Mimetic Theory," this volume.

8. Andrew Meltzoff and Keith Moore, "Imitation of Facial and Manual Gestures by Human Neonates," *Science* 198 (1977): 75–78; See also Andrew N. Meltzoff, "Out of the Mouths of Babes: Imitation, Gaze, and Intentions in Infant Research—the 'Like Me' Framework," in this volume; Colwyn Trevarthen, "Descriptive Analyses of Infant Communicative Behaviour," in *Studies in Mother-Infant Interaction*, ed. H. R. Schaffer (London: Academic Press, 1977); Colwyn Trevarthen, "Communication and Cooperation in Early Infancy: A Description of Primary Intersubjectivity," in *Before Speech: The Beginning of Interpersonal Communication*, ed. Margaret Bullowa (Cambridge: Cambridge University Press, 1979).

9. See Ann Cale Kruger, "Imitation, Communion, and Culture," this volume.

10. See Andrew N. Meltzoff, "Out of the Mouths of Babes: Imitation, Gaze, and Intentions in Infant Research—the 'Like Me' Framework," in this volume.

11. Beatrice Beebe, "Coconstructing Mother-Infant Distress: The Microsynchrony of Maternal Impingement and Infant Avoidance in the Face-to-Face Encounter," *Psychoanalytic Inquiry* 20 (2000): 421–440.

12. See Mark R. Anspach, "Imitation and Violence: Empirical Evidence and the Mimetic Model," this volume where he reports on an experimental study of infant rivalry for parental attention.

13. Michael Tomasello, *The Cultural Origins of Human Cognition* (Cambridge, MA: Harvard University Press, 2000). Merlin Donald, *Origins of the Modern Mind: Three Stages in the Evolution of Culture and Cognition* (Cambridge, MA: Harvard University Press, 1993).

14. Richard Dawkins, *The Selfish Gene* (Oxford: Oxford University Press, 1976).

15. Jane Goodall, *The Chimpanzees of Gombe: Patterns of Behavior* (Cambridge, MA: Harvard University Press, 1986).

16. For example, see Justin Barrett's *Why Would Anyone Believe in God?* (Walnut Creek, CA: AltaMira Press, 2004).

17. René Girard, *Battling to the End: Conversations with Benoît Chantre*, trans. Mary Baker (East Lansing: Michigan State University Press, 2010).

Contributors

Mark R. Anspach is an American anthropologist and social theorist based in Europe. He has been affiliated with the Centre de Recherche en Épistémologie Appliquée, École Polytechnique, Paris for more than twenty-five years. His research has focused on the ritual aspects of violence and exchange and on social and cognitive mechanisms of circular causality. He has contributed chapters in a number of books, including *Violence and the Sacred in the Modern World* (1991), *Introduction aux Sciences Cognitives* (1992), *Self-Deception and Paradoxes of Rationality* (1998), and *Che cos'è il religioso?* (2006), and he is the author of *Œdipe mimétique* (2010) and *À Charge de Revanche: Figures Élémentaires de la Réciprocité* (2002). He is also the editor of the *Cahier de L'Herne Girard* (2008) and of *Oedipus Unbound: Selected Writings on Rivalry and Desire* by René Girard (2004).

Paul Dumouchel is professor in the Graduate School of Core Ethics and Frontier Sciences at Ritsumeikan University where he teaches political philosophy and philosophy of science. He has published many articles on violence, war and terrorism, economics, the history of philosophy, and emotions as an object of research in both social and biological sciences. He is author of *Émotions essai sur le corps et le social* (1995) and co-author with Jean-Pierre Dupuy of *L'Enfer des Choses: René Girard et la logique de l'économie* (1979). He is also editor of *Violence and Truth: On the Work of René Girard* (1988) and *Comprendre*

Pour Agir: Violences, Victimes et Vengeances (2001). His most recent book is *Le Sacrifice inutile essai sur la violence et le politique* (2011).

Jean-Pierre Dupuy is Professor of French and, by courtesy, of Political Science, Stanford University, and Professor Emeritus of Social and Political Philosophy, Ecole Polytechnique, Paris. He is a member of the French Academy of Technology and of the Conseil Général des Mines, the French High Magistracy that oversees and regulates industry, energy and the environment. He chairs the Ethics Committee of the French High Authority on Nuclear Safety and Security. He is the Director of the Research Program of Imitatio, a new foundation devoted to the dissemination and discussion of René Girard's mimetic theory. His most recent work has dealt with the topic of catastrophe, and is being translated and collected in a volume to be published by Stanford University Press. His most recent publications include *The Mechanization of the Mind* (2000), *Pour un catastrophisme éclairé* (2002), *Avions-nous oublié le mal? Penser la politique après le 11 septembre* (2002), *Petite métaphysique des tsunamis* (2005), *Retour de Tchernobyl: Journal d'un homme en colère* (2006), *On the Origins of Cognitive Science* (2009), *La Marque du sacré* (2009), and *Dans l'œil du cyclone* (2009).

Vittorio Gallese is professor of physiology in the Department of Neuroscience of the School of Medicine at the University of Parma. He is renowned as a chief member of the team of researchers that discovered mirror neurons. He has worked at the University of Lausanne, Switzerland, and at the Nihon University of Tokyo, Japan, and in 2002 he was visiting professor at the Institute of Cognitive and Brain Sciences at the University of California at Berkeley. His major research interests are the relationship between action perception and cognition, using a variety of neurophysiological and neuro-imaging techniques. He is also interested in developing an interdisciplinary approach to the understanding of the embodied bases of intersubjectivity and social cognition. He has published approximately 100 papers in international peer-reviewed journals and books. He is also co-editor with Maxim Stamenov of *Mirror Neurons and the Evolution of Brain and Language* (2002).

Scott R. Garrels is adjunct professor in the School of Psychology at Fuller Theological Seminary and a licensed clinical psychologist in private practice in Pasadena, California. He earned his doctorate in clinical psychology from Fuller's Graduate School of Psychology where he investigated cognitive and psychosocial deficits in individuals with agenesis of the corpus callosum, a rare congenital brain abnormality. Over the last several years he has written and lectured widely on the interface between René Girard's mimetic theory of human violence, culture, and religion, and contemporary empirical research on human imitation, including the implications of these two fields of research for clinical theory and practice. As a result of his work, he received the Travis Award for Integration in Psychology and Theology from Fuller Theological Seminary (2003) and was awarded a Templeton Advanced Research Program grant from Metanexus Institute (2006).

René Girard is a member of the French Academy and emeritus professor of French Language, Literature, and Civilization at Stanford University. He is also the recipient of the Modern Language Association's Lifetime Achievement Award (2008). His work has inspired scholars from many diverse fields through his theories of human desire and of the significance of violence and religion in the formation of human culture. His many influential books have been translated widely and include *Deceit, Desire, and the Novel: Self and Other in Literary Structure* (1965), *Violence and the Sacred* (1977); *The Scapegoat* (1986), *Things Hidden Since the Foundation of the World* (1987), *A Theatre of Envy: William Shakespeare* (1991), *I See Satan Fall Like Lightening* (2001), *Oedipus Unbound: Selected Writings on Rivalry and Desire* (2004), and *Battling to the End: Conversations with Benoit Chantre* (2010).

William B. Hurlbut is a physician and consulting professor in the Department of Neurology and Neurological Sciences at Stanford University Medical Center. After receiving his undergraduate and medical training at Stanford, he completed postdoctoral studies in theology and medical ethics, studying with Robert Hamerton-Kelly, Dean of the Chapel at Stanford, and subsequently with Louis Bouyer of the Institut Catholique de Paris. In addition to teaching at Stanford, he served for eight years on the President's Council on Bioethics. His primary areas of interest involve the ethical issues associated with advancing biomedical technology, the biological basis of moral awareness, and studies in the integration of theology and philosophy of biology. He is the author of numerous publications on science and ethics, including the co-edited volume *Altruism and Altruistic Love: Science, Philosophy, and Religion in Dialogue* (2002), and "Science, Religion, and Human Spirit" in the *Oxford Handbook of Science and Religion* (2008).

Melvin Konner is Samuel Candler Dobbs Professor in the Department of Anthropology and the Program in Neuroscience and Behavioral Biology at

Emory University. He studied at Brooklyn College, The City University of New York, earned a doctorate in biological anthropology at Harvard, and did postdoctoral work at the Wurtman Laboratory of Neuroendocrine Regulation, Massachusetts Institute of Technology. He spent a total of two years doing fieldwork among the !Kung San (Bushmen), hunter-gatherers of Botswana, studying infant development and the hormonal mechanism of lactational infertility. After six years on the Harvard faculty, he attended Harvard Medical School and then moved to Emory University as department chair. He has held National Institute of Mental Health and National Science Foundation research grants, and has been a fellow of the Center for Advanced Study in the Behavioral Sciences at Stanford University, the John Simon Guggenheim Memorial Foundation, the Social Science Research Council, and the American Association for the Advancement of Science. He is the author of, among other books, *The Tangled Wing: Biological Constraints on the Human Spirit* (2nd edition, revised and updated 2002), *Becoming a Doctor* (1987), *Unsettled: An Anthropology of the Jews* (2003) and *The Evolution of Childhood: Relationships, Emotion, Mind* (2010).

Ann Cale Kruger is a developmental psychologist and associate professor of educational psychology at Georgia State University. Kruger earned her doctorate in developmental psychology from Emory University. Her research interests are in the areas of social cognition and cultural learning. She has written numerous articles and book chapters, several in collaboration with Michael Tomasello, addressing the role of joint attention and imitation for cultural learning, gestural communication, and object usage in both children and chimpanzees. Her current program of research focuses on how children use language and emotion understanding to learn from instruction and to collaborate. This focus has implications for the role of relationships, communication, and affect in instructional settings.

Andrew N. Meltzoff holds the Job and Gertrud Tamaki endowed chair and is the co-director of the University of Washington Institute for Learning and Brain Sciences. A graduate of Harvard University, with a doctorate from Oxford University, he is an internationally recognized expert on infant and child development. His discoveries about infant imitation have revolutionized our understanding of early cognition, personality, and brain development. Meltzoff's thirty years of research on young children has had far-reaching implications for cognitive science, especially for ideas about memory and its development; for brain science, especially for ideas about common coding and shared neural circuits for perception and action; and for early education and parenting, particularly for ideas about the importance of role models, both adults and peers, in child development. He is the co-author of two books about early learning and the brain: *The Scientist in the Crib: Minds, Brains, and How Children Learn* (1999) and *Words, Thoughts and Theories* (1997). He is also co-editor of *The Imitative Mind: Development, Evolution and Brain Bases* (2002).

Jean-Michel Oughourlian is professor of clinical psychopathology at the University of Paris. He taught at the Sorbonne for more than thirty years and served as chief of psychiatry at the American Hospital of Paris, where he is still in practice. For more than thirty years Oughourlian has worked in close collaboration with René Girard in their development of the mimetic theory of psychological mimesis and interdividual psychology. He was one of the primary collaborators with Girard on the book *Des Choses cachees depuis la fondation du monde* (*Things Hidden Since the Foundation of the World*, 1987). He is also the author of *Un mime nomme desir* (*The Puppet of Desire: The Psychology of Hysteria, Possession and Hypnosis* (1991) and, most recently, *Genese du desir* (*The Genesis of Desire*, 2010).

Index